What Went Wrong?

Case Histories of Process Plant Disasters

Gulf Publishing Company
Book Division
Houston, London, Paris, Tokyo

What Went Wrong?

Case Histories of Process Plant Disasters

Trevor A. Kletz

*To Denise who waited while I "scorned delights, and
lived laborious days," but never saw the results.*

What Went Wrong?

Case Histories of Process Plant Disasters

Library of Congress Cataloging in Publication Data

Kletz, Trevor A.
What went wrong?

Includes index.
1. Chemical plants—Accidents. I. Title.
TP155.5.K54 1985   363.1'1966   84-23483

ISBN 0-87201-339-1

Contents

Acknowledgments

Thanks are due to the staffs of the companies where the incidents occurred for allowing me to describe their mistakes; to many colleagues, past and present; and especially to Professor F. P. Lees for his ideas and advice. To the UK Science and Engineering Research Council for its financial support.

I shall be pleased to hear of other incidents to be included in a future volume. Write to T. A. Kletz, Department of Chemical Engineering, University of Technology, Loughborough, Leicestershire, LE11 3TU, England.

Preface

In 1968, after many years' experience in plant operations, I was appointed safety adviser to the Heavy Organic Chemicals Division (later the Petrochemicals Division) of Imperial Chemical Industries Ltd. My appointment followed a number of serious fires in the 1960s and therefore I was mainly concerned with process hazards rather than with those of a mechanical nature.

One of my tasks was to pass on to design and operating staff details of accidents that had occurred and the lessons that should be learned. This book contains a selection of the reports I collected from many different companies. Although most of them have been published before, they were scattered among many different publications, some of small circulation.

The purpose here is to show what has gone wrong in the past and to suggest how similar incidents might be prevented in the future. Unfortunately, the history of the process industries shows that many incidents are repeated after a lapse of a few years*.

People move and the lessons are forgotten. This book will help to keep the memories alive.

The advice is given in good faith but without warranty. Readers should satisfy themselves that it is applicable to their circumstances. In fact, you may feel that some of my recommendations are not appropriate for your company. Fair enough, but if the incidents could occur in your company,

---

* See T.A. Kletz, *Loss Prevention Series,* AIChE, Vol. 10, 1976, p. 51, and Vol. 13, 1980, p. 1.

and you do not wish to adopt my advice, then please do something else instead. But do not ignore the incidents.

The incidents described could occur in many different plants and are therefore of widespread interest. Some of them illustrate the hazards involved in activities such as preparing equipment for maintenance or modifying plants. Others illustrate the hazards associated with widely used equipment such as storage tanks or hoses or with that universal component of all plants and processes: people. Other incidents illustrate the need for techniques such as hazard and operability studies or protective devices such as emergency isolation valves.

You will notice that most of the incidents are very simple. No esoteric knowledge or detailed study was required to prevent them—only a knowledge of what has happened before, which this book provides.

Many of the incidents described could be discussed under more than one heading. Cross-references have therefore been included.

If an incident that happened in your plant is described, you may notice that one or two details have been changed. Sometimes this has been done to make it harder for people to tell where the incident occurred. Sometimes this has been done to make a complicated story simpler, but without affecting the essential message. Sometimes—and this is the most likely reason—the incident did not happen in your plant at all. Another plant had a similar incident.

Many of the incidents did not actually result in death, serious injury or serious damage—they were near-misses. But they could have had much more serious consequences. We should learn from these near-misses as well as from incidents that had serious results.

Most of the incidents described occurred at so-called major-hazard plants or storage installations—that is, those containing large quantities of flammable, explosive, or toxic chemicals. The lessons learned apply particularly to such plants.

However, most of the incidents could have occurred at plants handling smaller quantities of materials or less hazardous materials and the consequences, though less serious, would be serious enough. At a major-hazard plant, opening up a pump which is not isolated could cause (and has caused) a major fire or explosion. At other plants it would cause a smaller fire, or a release of corrosive chemicals, still enough to kill or injure the employee on the job.

Even if the contents of the plant are harmless, there is still a waste of materials.

The lessons to be learned therefore apply throughout the process industries.

HOW TO USE THIS BOOK

1. Read it right through. As you do so, ask yourself if the incidents could occur in *your* plant, and, if so, *write down* what you intend to do to prevent them from occurring.
2. Use it as a deskside book on safety. Dip into it at odd moments, or pick a subject for the staff meeting, the safety committee or bulletin, or the plant inspection.
3. Refer to it when you become interested in something new, as the result of an incident, a change in responsibility, or a new problem in design. However, this book does not claim to be a comprehensive review of process safety and loss prevention. For that, refer to F.P. Lees' *Loss Prevention in the Process Industries,* Butterworths, 1980 (2 volumes).
4. Use the incidents to train new staff, managers, foremen and operators, so that they know what will happen if they do not follow recognized procedures and good operating practice.
5. If you are a teacher, use the incidents to tell your students why accidents occur and use the incidents to illustrate scientific principles.

   Both in the training of plant staff and students, the material can be used as lecture material or, better, as discussion material (those present discuss and agree among themselves what they think should be done to prevent similar incidents happening again). The use of case histories in this way is discussed in *Health and Safety at Work,* Vol. 1, 1978, p. 41, and Vol. 3, 1981, p. 40, by D. A. Lihou, and *An Introduction to Hazard Workshop Training Modules,* by D. A. Lihou and G. S. G. Beveridge (Institution of Chemical Engineers, 1981).
6. If you want to be nasty, send a copy of the book, open at the appropriate page, to people who have allowed one of the accidents described to happen again. They may read the book and avoid further unnecessary accidents.

A high price has been paid for the information in this book: Many persons killed and billions of dollars worth of equipment destroyed. You get this information for the price of the book. It will be the best bargain you have ever had, if you use the information to prevent similar incidents at your plant.

Trevor A. Kletz

Units and Nomenclature

I have used units likely to be most familiar to the majority of my readers. Although I welcome the increasing use of SI units, many people still use imperial units—they are more familiar with a 1-inch pipe than a 25 millimeter pipe.

Short lengths are therefore quoted in inches but longer lengths in meters.

1 inch = 25.4 mm

1 m   =   3.28 feet or 1.09 yards

Volumes are quoted in cubic meters (m^3) as this unit is widely used and "gallons" is ambiguous.

$1 m^3$ = 264 US gallons
       = 220 imperial gallons
       = 35.3 cubic feet

Masses are quoted in kilograms (kg) and tons (t)

1 kg =   2.20 pounds

1 t   =   1.10 short (US) tons
      =   0.98 long (UK) ton

Temperatures are quoted in °C

Pressures are quoted in pounds force per square inch (psi) and also in bars. As it is not usual to refer to "bar gauge" I have, for example, referred to "a gauge pressure of 90 psi (6 bar)," rather than "a pressure of 90 psig."

$$1 \text{ bar} = 14.50 \text{ psi}$$
$$= 1 \text{ atmosphere}$$
$$= 1 \text{ kilogram per square cm}$$
$$= 100 \text{ kilopascals (kPa)}$$

Very small gauge pressures are quoted in inches water gauge as this gives a picture.

$$1 \text{ inch water gauge} = 0.036 \text{ psi}$$
$$= 2.5 \times 10^{-3} \text{ bar}$$
$$= 0.2 \text{ kPa}$$

A NOTE ON NOMENCLATURE

Different words are used, in different countries, to describe the same job or piece of equipment. Some of the principal differences between the U.S. and the UK are listed here. Within each country, however, there are differences between companies.

MANAGEMENT TERMS

Job	US	UK
Operator of plant	Operator	Process worker
Operator in charge of others	Lead operator	Chargehand or Assistant Foreman or Junior Supervisor
Highest level normally reached by promotion from operator	Foreman	Foreman or Supervisor
First level of professional management (usually in charge of a single unit)	Supervisor	Plant manager
Second level of professional management	Superintendent	Section manager
Senior Manager in charge of site containing many units.	Plant Manager	Works manager
	Craftsman or mechanic	Fitter, electrician, etc.

The different meanings of the terms "supervisor" and "plant manager" in the U.S. and UK should be noted.

In this book I have used the term "foreman" as it is understood in both countries, though its use in the UK is now becoming outdated. "Manager" is used to describe any professionally qualified person in charge of a unit or group of units. That is, it includes people who, in many U.S. companies, would be described as supervisors or superintendents.

Certain items of plant equipment have different names in the two countries. Some common examples are:

CHEMICAL ENGINEERING TERMS

US	UK
Agitator	Mixer or stirrer
Blind	Slip-plate
Carbon steel	Mild steel
Carrier	Refrigeration plant
Check valve	Nonreturn valve
Clogged (of filter)	Blinded
Consensus standard	Code of practice
Conservation vent	Pressure/vacuum valve
Dike	Bund
Division (in electrical area classification)	Zone
Downspout	Downcomer
Fiberglass reinforced plastic (FRP)	Glass reinforced plastic (GRP)
Figure 8 plate	Spectacle plate
Flame arrestor	Flame trap
Flashlight	Torch
Forcet	Tap
Fractionation	Distillation
Gaging (of tanks)	Dipping
Gasoline	Petrol
Generator	Dynamo or alternator
Ground	Earth
Hydro (Canada)	Electricity
Install	Fit
Insulation	Lagging
Inventory	Stock

US	UK
Interlock	Trip*
Lift-truck	Fork lift truck
Manway	Manhole
Mill water	Cooling water
Nozzle	Branch
Pedestal, pier	Plinth
Rupture disc or frangible	Bursting disc
Scrutinize	Vet
Shutdown	Permanent shutdown
Sieve tray	Perforated plate
Siphon tube	Dip tube
Spade	Slip-plate
Sparger or sparge pump	Spray nozzle
Spigot	Tap
Stack	Chimney
Stator	Armature
Tank car	Rail tanker or rail tank wagon
Tank truck	Road tanker or road tank wagon
Torch	Cutting or welding torch
Tower	Column
Tow motor	Fork lift truck
Tray	Plate
Turnaround	Shutdown
Utility hole	Manhole
Wrench	Spanner
C-wrench	Adjustable spanner
$M	Thousand dollars
$MM	$M or million dollars
STP	60°F, 1 atmosphere
32°F, 1 atmosphere	STP
NTP	32°F, 1 atmosphere

---

* In the UK "interlock" is used to describe a device which prevents someone opening one valve while another is open (or closed). "Trip" describes an automatic device which closes (or opens) a valve when a temperature, pressure, flow, etc. reaches a pre-set value.

FIRE-FIGHTING TERMS

US	*UK*
Dry chemical	Dry powder
Dry powder	Dry powder for metal fires
Excelsior (for fire tests)	Wood wool
Egress	Escape
Sprinkler systems:	
Feed main	Main distribution pipe
Cross main	Distribution pipe
Branch pipe	Range pipe
Rate density	Application rate
Fire stream	Jet
Standpipe	Dry riser
Evolutions	Drills
Nozzle	Branchpipe
Tip	Nozzle
Siamese connection	Collecting breeching
Wye connection	Dividing breeching
Open butt	Hose without branchpipe
Fire classification:	
Class A: Solids	Class A: Solids
Class B: Liquids and gases	Class B: Liquids
Class C: Electrical	Class C: Gases
Class D: Metals	Class D: Metals

Chapter 1

Preparation for Maintenance

The following pages describe accidents which occurred because equipment was not adequately prepared for maintenance. Sometimes it was not isolated from hazardous materials; sometimes it was not identified correctly so that the wrong equipment was opened up; sometimes hazardous materials were not removed (1, 2).*

Entry to vessels is discussed in Chapter 11.

1.1 ISOLATION

1.1.1 FAILURE TO ISOLATE

A pump was being dismantled for repair. When the cover was removed, hot oil, above its auto-ignition temperature, came out and caught fire. Three men were killed and the plant was destroyed. Examination of the wreckage after the fire showed that the pump suction valve was open and the drain valve shut (3).

The pump had been awaiting repair for several days when a permit-to-work was issued at 8 a.m. on the day of the fire. The foreman who issued the permit should have checked, before doing so, that the pump suction and delivery valves were shut and the drain valve open. He claimed that he did so. Either his recollection was incorrect or, after he inspected the valves and before work started, someone closed the drain valve and opened the suction valve. When the valves were closed, there was no indication—on them—of *why* they were closed. An operator might have opened the suction valve and shut the drain valve so that the pump could be put on line quickly if required.

---

* End of chapter references are indicated by a number inside parentheses.

A complicating factor was that the maintenance team originally intended to work only on the pump bearings. When they found that they had to open up the pump they told the process team, but no further checks of the isolations were carried out.

It was not customary in the company concerned to isolate equipment under repair by slip-plates, only by closed valves. But after the fire the following rules were introduced:

(a) Equipment under repair must be isolated by slip-plates (blinds or spades) or physical disconnection unless the job to be done is so quick that fitting slip-plates (or disconnecting pipework) would take as long as the main job and be as hazardous. If hot work is to be carried out or a vessel is to be entered, then slip-plating or physical disconnection must always take place.

(b) Valves isolating equipment under maintenance, including valves which have to be closed while slip-plates are fitted (or pipework disconnected), must be locked shut with a padlock and chain or similar device. A notice fixed to the valve is not sufficient.

(c) For fluids at gauge pressures above 600 psi (40 bar) or at a temperature near or above the auto-ignition point, double block and bleed valves should be installed—not for use as main isolations, but so that slip-plates can be inserted safely (Figure 1.1).

(d) If there is any change in the work to be done the permit-to-work must be withdrawn and a new one issued.

Another similar incident is described in Section 18.1.

1.1.2 ISOLATIONS REMOVED TOO SOON

An ethylene compressor was shut down for maintenance and correctly isolated by slip-plates. When repairs were complete the slip-plates were removed before the machine was tried out. During the try-out some ethylene leaked through the closed isolation valves into the machine. The ethylene/air mixture was ignited, either by a hot spot in the machine or by copper acetylide on the copper valve gaskets. The compressor was severely damaged.

Isolations should not be removed until maintenance is complete. It is best to issue three work permits—one for inserting slip-plates (or disconnecting pipework), one for the main job and one for removing slip-plates (or restoring disconnections).

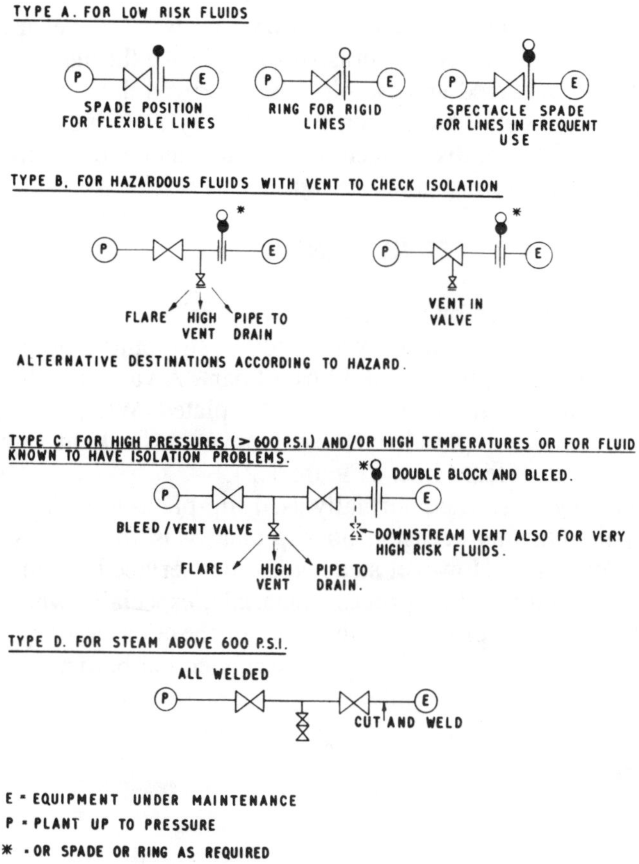

TYPE A. FOR LOW RISK FLUIDS

SPADE POSITION
FOR FLEXIBLE LINES RING FOR RIGID
LINES SPECTACLE SPADE
FOR LINES IN FREQUENT
USE

TYPE B. FOR HAZARDOUS FLUIDS WITH VENT TO CHECK ISOLATION

FLARE HIGH PIPE TO
VENT DRAIN VENT IN
VALVE

ALTERNATIVE DESTINATIONS ACCORDING TO HAZARD.

TYPE C. FOR HIGH PRESSURES (> 600 P.S.I.) AND/OR HIGH TEMPERATURES OR FOR FLUID
KNOWN TO HAVE ISOLATION PROBLEMS.

DOUBLE BLOCK AND BLEED.

BLEED / VENT VALVE

DOWNSTREAM VENT ALSO FOR VERY
HIGH RISK FLUIDS.

FLARE HIGH PIPE TO
VENT DRAIN.

TYPE D. FOR STEAM ABOVE 600 P.S.I.

ALL WELDED

CUT AND WELD

E - EQUIPMENT UNDER MAINTENANCE
P - PLANT UP TO PRESSURE
* - OR SPADE OR RING AS REQUIRED

Figure 1.1. Summary of isolation methods.

1.1.3 INADEQUATE ISOLATION

A reactor was prepared for maintenance and washed out. There was no welding to be done and no entry was required, so it was decided not to slip-plate off the reactor but to rely on valve isolations. Some flammable vapor leaked through the closed valves into the reactor and was ignited by a high speed abrasive wheel which was being used to cut through one of the pipe-lines attached to the vessel. The reactor head was blown off and killed two men. It was estimated that 7 kg of hydrocarbon vapor could have caused the explosion.

Following the accident, demonstration cuts were made in the workshop. It was found that as the abrasive wheel broke through the pipe wall a small flame occurred and the pipe itself glowed dull red.

The explosion could have been prevented by isolating the reactor by slip-plates or physical disconnection. This incident and the previous one show that valves are not good enough.

1.1.4 ISOLATION OF SERVICE LINES

A fitter was affected by fumes while working on a steam drum. One of the steam lines from the drum was used for stripping a process column operating at a gauge pressure of 30 psi (2 bar). A valve on the line to the column was closed but the line was not slip-plated. When the steam pressure was blown off, vapors from the column came back through the leaking valve into the steam lines (Figure 1.2).

The company concerned normally used slip-plates to isolate equipment under repair. On this occasion no slip-plate was fitted because it was "only" a steam line. However steam and other service lines in plant areas are easily contaminated by process materials, especially when there is a direct connection to process equipment. In these cases, they should be positively isolated by slip-plating or disconnection before maintenance.

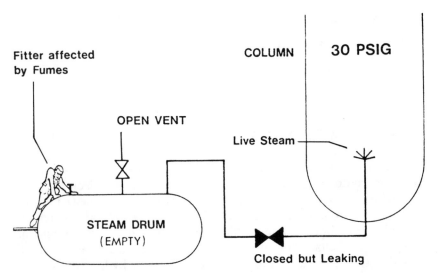

Figure 1.2. Contamination of a steam drum by process materials.

1.1.5 ISOLATIONS NOT REMOVED

While a plant was on line an operator noticed a slip-plate on a tank vent. The slip-plate had been fitted to isolate the tank from the blow-down system while the tank was under maintenance. When the maintenance was complete, the slip-plate was overlooked.

Fortunately, the tank, an old one, was stronger than it needed to be for the duty, or it would have burst.

If a vessel has to be isolated from the vent or blow-down line, do not slip-plate it off but, whenever possible, disconnect it and leave the vessel vented to atmosphere, as shown in Figure 1.3.

If the vent line forms part of a blow-down system it will have to be blanked to prevent air being sucked in. Make sure the blank is put on the flare side of the disconnection, not on the tank side (Figure 1.3).

Note that if the tank is to be entered, the joint nearest the tank should be broken.

If a vent line has to be slip-plated, because the line is too rigid to be moved, then the vents should be slip-plated last and de-slip-plated first.

If all slip-plates inserted are listed on a register, they are less likely to be overlooked.

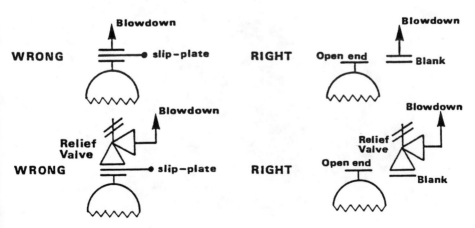

Figure 1.3. The right and wrong ways to isolate a vent line.

1.1.6 SOME MISCELLANEOUS INCIDENTS INVOLVING ISOLATION FOR MAINTENANCE

(a) A slip-plate that had been in position for many months, perhaps years, was relied on to isolate equipment. It had corroded right through (Figure 1.4).

Slip-plates in position for a long time should be removed and inspected before being used as maintenance isolations. (Such slip-plates should be registered for inspection every few years.)

(b) A slip-plate with a short tag was overlooked and left in position when maintenance was complete.

Tags should be at least 130 mm long, on lines up to and including 6 inches diameter and at least 150 mm long on larger lines.

(c) On several occasions small bore branches have been covered by insulation, overlooked and not isolated.

(d) On several occasions thin slip-plates have been used and have become bowed; they are then difficult to remove. Figure 1.5 shows a thin slip-plate which has been subjected to a gauge pressure of 470 psi (32 bar).

Slip-plates should normally be designed to withstand the same pressure as the piping. However, in some older plants which have not been designed to take full-thickness slip-plates it may be impossible to insert them. A compromise will be necessary.

Figure 1.4. A slip-plate left in position for many months had corroded right through.

Figure 1.5. A slip-plate bowed by a gauge pressure of 470 psi (32 bar).

1.2 IDENTIFICATION

1.2.1 THE NEED FOR TAGGING

On many occasions the wrong pipeline or piece of equipment has been broken into. For example:

(a) A joint which had to be broken was marked with chalk. The fitter broke another joint which had an old chalk mark on it. He was splashed with a corrosive chemical.

(b) An out-of-service pipeline was marked with chalk at the point at which it was to be cut. Before the fitter could start work, a heavy rain washed off the chalk mark. The fitter "remembered" where the chalk mark had been. He was found cutting his way with a hacksaw through a line containing a hazardous chemical.

(c) Water was dripping from a joint on a line on a pipebridge. Scaffolding was erected to provide access for repair. But to avoid having to climb up onto the scaffold the process foreman pointed out the leaking joint from the ground and asked a fitter to remake the joint in the "water line." The joint was actually in a carbon monoxide line. So, when the fitter broke the joint he was overcome and, because of the poor access, was rescued only with difficulty.

If the process foreman had gone up to the joint on the pipebridge he would have realized that the water was dripping out of the carbon monoxide line.

(d) The bonnet had to be removed from a steam valve. It was pointed out to the fitter from the floor above. He went down a flight of stairs, approached the valve from the side and removed the bonnet from a compressed air valve. It flew off, grazing his face.

All these incidents and many more could be prevented by fitting a numbered tag to the joint or valve and putting the number on the work permit. In incident (c) the foreman would have had to go up onto the scaffold to fix the tag.

Accidents have occurred, however, despite tagging systems.

In one plant a fitter did not check the tag number and broke a joint which had been tagged for an earlier job; the tag had been left in position. Tags should be removed when jobs are complete.

In another plant the foreman allowed a planner to fix the tags for him and did not check that they were fixed to the right equipment. The foreman prepared one line for maintenance but the tags were on another.

1.2.2 THE NEED FOR CLEAR, UNAMBIGUOUS LABELING

(a) A row of pumps were labeled as shown in Figure 1.6. A fitter was asked to repair number 7. Not unreasonably he assumed that number 7 was the end one. He did not check the numbers. Hot oil came out of the pump when he dismantled it.

(b) There were four crystallizers in a plant, three old ones and one just installed. A man was asked to repair A. When he went onto the structure he saw that two were labeled B and C but the other two were not labeled. He assumed that A was the old unlabeled crystallizer and started work on it. Actually A was the new crystallizer. The original three were called B, C and D. A was reserved for a possible future addition for which space was left (Figure 1.7).

(c) The labels on two air coolers were arranged as shown in Figure 1.8. The B label was on the side of the B cooler furthest away from

Figure 1.6. Numbering pumps like this leads to error.

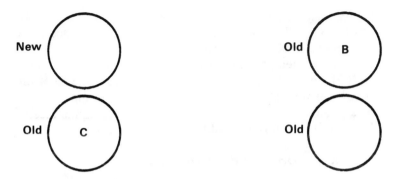

Figure 1.7. Which is "A" crystallizer?

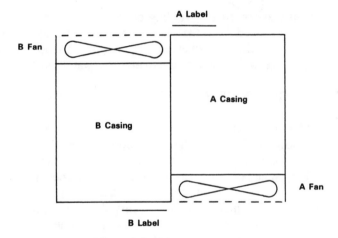

Figure 1.8. Which is "A" fan?

the fan and near A fan. Not unreasonably, workers asked to over-haul B fan assumed it was the one next to the B label and over-hauled it. The power had not been isolated. But fortunately the overhaul was nearly complete before someone started the fan.

(d) Some pump numbers were painted on the coupling guards. Before long, repairs were carried out on the couplings of two adjacent pumps. You can guess what happened. Now the pump numbers are painted on the pump bodies. It would be even better to paint the numbers on the plinths.

(e) On one unit the pumps and compressors were numbered J1001 on-wards. When the unit's allocation of numbers was used up, num-bers from JA1001 onwards were used. J1001 and JA1001 *sound* alike (say them aloud). An operator was asked to prepare JA1001—a small pump—for repair. He thought the foreman said J1001 and went to it. J1001 was a 40,000 HP compressor. Fortu-nately, the size of the machine made him hesitate. He asked the foreman if he really wanted the compressor shut down.

1.2.3 THE NEED FOR CLEAR INSTRUCTIONS

(a) A permit was issued for modifications to the walls of a room. The maintenance workers started work on the ceilings as well and cut through live electric cables.
(b) A permit was issued for welding on the *top only* of a tank which had been removed from the plant. When the job was complete the welders rolled the tank over so that another part became the top. Some residue which had been covered by water caught fire.

1.2.4 IDENTIFICATION OF RELIEF VALVES

Two relief valves, identical in appearance, were removed from a plant during a shutdown and sent to the workshops for overhaul. One relief valve was set to operate at a gauge pressure of 15 psi (1 bar) and the other at 30 psi (2 bar). The set pressures were stamped on the flanges but this did not prevent the valves being interchanged.

A number of similar incidents have occurred in other plants.

Such incidents can be prevented, or at least made much less likely, by tying a numbered tag to the relief valve when it is removed and tying another tag with the same number to the flange.

1.3 REMOVAL OF HAZARDS

Many accidents have occurred because equipment, though isolated correctly, was not completely freed from hazardous materials or because the pressure inside it was not completely blown off and *the workers car-rying out the repair were not made aware of this.*

1.3.1 EQUIPMENT NOT GAS FREED

It is usual to test for the presence of flammable gas or vapor with a combustible gas detector before maintenance, especially welding or other hot work, is allowed to start.

The following incidents show what can happen if these tests are not carried out.

(a) An explosion occurred in a 4,000 m^3 underground storage tank at Sheffield Gas Works, England, in October 1973. Six people were killed, 29 injured and the tank was destroyed. The tank top was thrown into the air, turned over and deposited upside-down on the bottom of the tank.

The tank had contained a light naphtha and had not been thoroughly cleaned before repairs started. It had been filled with water and then emptied but some naphtha remained in various nooks and crannies. (It might, for example, have gotten into the hollow roof supports through pin-holes or cracks, and then drained out when the tank was emptied.) No tests were carried out with combustible gas detectors.

It is believed that the vapor was ignited by welding near an open vent. The body of the welder was found 100 feet up on the top of a neighboring gasholder still holding a welding torch.

According to the incident report there was no clear division of responsibilities between the Gas Board and the contractor who was carrying out the repairs.

"Where, as in this case, a special risk is likely to arise due to the nature of the work performed (and the owner of the premises has special knowledge of it), the owner must retain sufficient control of the operation to ensure that contractors' employees are properly protected against the risk (4)."

(b) A bottom manhole was removed from an empty tank still full of gasoline vapor. Vapor came out of the manhole and caught fire. As the vapor burned, air was sucked into the tank through the vent until the contents became explosive. The tank then blew up (5).

(c) Welding had to be carried out—during a shutdown—on a relief valve tailpipe. It was disconnected at both ends. Four hours later the atmosphere at the end furthest from the relief valve was tested with a combustible gas detector. The head of the detector was pushed as far down the tailpipe as it would go; no gas was detected and a work permit was issued. While the relief valve discharge flange was being ground, a flash and bang occurred at the other end of the tailpipe. Fortunately, no one was hurt. Gas in the tailpipe—20 m long and containing a number of bends—had not dispersed and had not been detected by a test at the other end of the pipe.

Before allowing welding or similar operations on a pipeline which has or could have contained flammable gas or liquid, (1) sweep out the line with steam or nitrogen from end to end, and (2) test at the point at which welding will be carried out. If necessary, a hole may have to be drilled in the pipeline.

1.3.2 CONDITIONS CAN CHANGE AFTER TESTING

As already stated, it is usual to test for the presence of flammable gas or vapor with a combustible gas detector before maintenance, especially welding or other hot work, starts. Several incidents have occurred because tests were carried out several hours beforehand and conditions changed.

(a) An old propylene line which had been out of use for 12 years had to be modified for reuse. For the last two years it had been open at one end and blanked at the other. The first job was welding a flange onto the open end. This was done without incident. The second job was to fit a 1-inch branch 60 m from the open end. A hole was drilled in the pipe and the inside of the line tested. No gas was detected. Fortunately, a few hours later, just before welding was about to start, the inside of the pipe was tested again and flammable gas was detected. It is believed that some gas had remained in the line for 12 years and a slight rise in temperature had caused it to move along the pipeline. Some people might consider that a line out of use for 12 years did not need testing at all. Fortunately the men concerned did not take this view. They tested the inside of the line and tested again immediately before welding started.

(b) A test for benzene in the atmosphere was carried out 8 hours before a job started. During this time the concentration of benzene rose.

(c) An acid tank was prepared for welding and a permit issued. It was *40 days* before the maintenance team was able to start. During this time a small amount of acid which had been left in the tank attacked the metal, producing hydrogen. No further tests were carried out. When welding started, an explosion occurred (6).

(d) A branch had to be welded onto a pipeline which was close to the ground. A small excavation, between 1/2 and 1 m deep, was made to provide access to the bottom of the pipeline. The atmosphere in the excavation was tested with a combustible gas detector and because no gas was detected a welding permit was issued. Half an hour later, after the welder had started work, a small fire occurred in the excavation. Some hydrocarbons had leaked out of the ground. This incident shows that it may not be sufficient to test just before welding starts. It may be necessary to carry out continuous tests using a portable combustible gas detector alarm.

1.3.3 VAPOR CAN COME OUT OF DRAINS AND VENTS

A number of incidents have occurred because gas or vapor came out of drains or vents while work was in progress. For example:

(a) Welding had to be carried out on a pipeline 20 feet above the ground. Tests inside and near the pipeline were negative and so a work permit was issued. A piece of hot welding slag bounced off a pipeline and fell onto a sump 6 m below and 2.5 m to the side. The cover on the sump was loose and some oil inside caught fire. Welding jobs should be boxed in with fire-resistant sheets. Nevertheless, some sparks or pieces of slag may reach the ground. So drains and sumps should be covered.

(b) While an electrician was installing a new light on the outside wall of a building, he was affected by fumes coming out of a ventilation duct 0.6 m away. When the job was planned, the electrical hazards were considered and also the hazards of working on ladders. But it did not occur to anyone that harmful or unpleasant fumes might come out of the duct. Yet ventilation systems are installed to get rid of fumes.

1.3.4 LIQUID CAN BE LEFT IN LINES

When a line is drained or blown clear, liquid may be left in low-lying sections and run out when the line is broken. This is particularly hazardous if overhead lines have to be broken. Liquid splashes down onto the ground. Funnels and flexes should be used to catch spillages.

When possible, drain points in a pipeline should be fitted at low points and slip-plates should be fitted at high points.

1.3.5 SERVICE LINES MAY CONTAIN HAZARDOUS MATERIALS

Section 1.1.4 described how fumes got into a steam drum because it was not properly isolated. Even when service lines are not directly connected to process materials they should always be tested before maintenance, particularly if hot work is permitted on them, as the following incidents show:

(a) A steam line was blown down and cold cut. Then a plug was hammered into one of the open ends. A welder struck an arc ready to weld in the plug. An explosion occurred and the plug was blown out of the pipeline, fortunately missing the welder. Acid had leaked into the pipeline through a corroded heating coil in an acid tank and had reacted with the iron of the steam pipe, producing hydrogen.

(b) While a welder was working on the water line leading to a waste heat boiler, gas came out of a broken joint and caught fire. The welder was burned, but not seriously. There was a leaking tube in the waste heat boiler. Normally, water leaked into the process stream. However, on shutting down the plant, pressure was taken off the water side before it was taken off the process side, thus reversing the leak direction. The water side should have been kept up to pressure until the process side was depressured. In addition, the inside of the water lines should have been tested with a combustible gas detector.

See also Section 5.4.2 (b).

1.3.6 TRAPPED PRESSURE

Even though equipment is isolated by slip-plates and the pressure has been blown off through valves or by cracking a joint, pressure may still be trapped elsewhere in the equipment, as the following incidents show:

(a) This incident occurred on an all-welded line. The valves were welded in. To clear a choke, a fitter removed the bonnet and inside of a valve. He saw that the seat was choked with solid and started to chip it away. As he did so, a jet of corrosive chemical came out under pressure from behind the solid, hit him in the face, pushed his goggles aside and entered his eye.

(b) An old acid line was being dismantled. The first joint was opened without trouble. But when the second joint was opened acid came out under pressure and splashed the fitter and his assistant in the face. Acid had attacked the pipe, building up gas pressure in some parts and blocking it with sludge in others.

(c) A joint on an acid line, known to be choked, was carefully broken, but only a trickle of acid came out. More bolts were removed and the joint pulled apart, but no more acid came. When the last bolt was removed and the joint pulled wide apart a sudden burst of pressure blew acid into the fitter's face.

In all three cases the lines were correctly isolated from operating equipment. Work permits specified that goggles should be worn and stated "Beware of trapped pressure."

To avoid injuries of this sort we should use protective hoods or helmets when breaking joints on lines likely to contain corrosive liquids trapped under pressure, either because the pressure cannot be blown off through a valve or because lines may contain solid deposits.

Other incidents due to trapped pressure are described in Section 17.1.

1.3.7 EQUIPMENT SENT OUTSIDE THE PLANT

When a piece of equipment is sent to a workshop or to another company for repair or modification we should, whenever possible, make sure that it is spotlessly clean before it leaves the plant. Contractors are usually not familiar with chemicals and do not know how to handle them.

Occasionally, however, it may be impossible to be certain that piece of equipment is spotlessly clean, especially if it has contained a residual oil or a material which polymerizes. If this is the case, or if there is some doubt about its cleanliness, then the hazards and the necessary precautions should be made known to the workshop or the other company. This can be done by attaching a certificate to the equipment. This certificate is not a work permit. It does not authorize any work, but describes the state of the equipment and gives the other company sufficient information to enable them to carry out the repair or modification safely. Before issuing the certificate the engineer in charge should discuss with the other company the methods they propose to use. If the problems are complex, a member of the plant staff may have to visit the other company. The following incident shows the need for these precautions.

A large heat exchanger, 8 feet long by 8½ feet diameter, was sent to another company for retubing. It contained about 800 2½-inch-diameter tubes, of which about 80 had been plugged. The tubes had contained a process material which tends to form chokes, and the shell had contained steam.

Before the exchanger left the plant the free tubes were cleaned with high pressure water jets. The plugged tubes were opened up by drilling ³/₈-inch holes through the plugs to relieve any trapped pressure. But these holes were not big enough to allow the tubes to be cleaned.

A certificate was attached to the exchanger stating that welding and burning were allowed, but only to the shell.

The contractor, having removed most of the tubes, decided to put workers into the shell to grind out the plugged tubes. He telephoned the plant and asked if it would be safe to let workers enter the shell. He did not say why he wanted them to do so.

The plant engineer who took the telephone call said that the shell side was clean and therefore it would be safe to enter it. He was not told that the workers were going into it to grind out some of the tubes.

Two men went into the shell and started grinding. They were affected by fumes and the job was left to the next day. Another three workers then restarted the job and were affected so badly that they were hospitalized. Fortunately, they soon recovered.

The certificate attached to the exchanger when it left the plant should have contained much more information. It should have said that the plugged tubes had not been cleaned and that they contained a chemical which gave off fumes when heated. Better still, the plugged tubes should have been opened up and cleaned. The contractor would have to remove the plugs, so why not remove them before they left the plant?

Do your instructions cover the points mentioned in this section?

1.4 PROCEDURES NOT FOLLOWED

It is usual, before a piece of equipment is maintained, to give the maintenance team a permit-to-work which sets out:

1. What is to be done.
2. How the equipment is isolated and identified.
3. What hazards, if any, remain.
4. What precautions should be taken.

This section describes incidents which occurred because of loopholes in the procedure for issuing work permits or because the procedure was not followed. There is no clear distinction between these two categories. Often the procedure does not cover, or seem to cover, all circumstances. Those concerned use this as the reason, or excuse, for a shortcut, as in the following two incidents:

1.4.1 EQUIPMENT USED AFTER A PERMIT HAS BEEN ISSUED

(a) A plumber foreman was given a work permit to modify a pipeline. At 4 p.m. the plumbers went home, intending to complete the job on the following day. During the evening the process foreman wanted to use the line the plumbers were working on. He checked that the line was safe to use and he asked the shift maintenance man to sign off the permit. Next morning the plumbers, not knowing that their permit had been withdrawn, started work on the line while it was in use.

To prevent similar incidents happening again: (1) It should be made clear that permits can only be signed off by the person who has accepted them (or a person who has taken over his responsibilities), and (2) there should be two copies of every permit, one kept by the maintenance team and one left in the book in the possession of the process team.

(b) A manhole cover was removed from a reactor so that some extra catalyst could be put in. After the cover had been removed, it was found that the necessary manpower would not be available until the next day. So it was decided to replace the manhole cover and re-generate the catalyst overnight. By this time it was evening and the maintenance foreman had gone home and left the work permit in his office, which was locked. The reactor was therefore boxed up and catalyst regeneration carried out with the permit still in force.

The next day a fitter, armed with the work permit, proceeded to remove the manhole cover again, and while doing so was drenched with process liquid. Fortunately, the liquid was mostly water and he was not injured.

The reactor should not have been boxed up and put on line until the original permit had been handed back. If it was locked up, then the maintenance supervisor should have been called in. Except in an emergency, plant operations should never be carried out while a work permit is in force on the equipment concerned.

1.4.2 PROTECTIVE CLOTHING NOT WORN

The following incidents are typical of many.

(a) A permit issued for work to be carried out on an acid line stated that goggles must be worn. Although the line had been drained, there might have been some trapped pressure (see Section 1.3.6). The man doing the job did not wear goggles and was splashed in the eye.

At first sight, the injury was entirely the fault of the injured man and there was nothing that anyone else could have done to prevent it. However, further investigation showed that *all* permits issued asked for goggles to be worn, even for repairs to water lines. The maintenance workers therefore frequently ignored this instruction and the managers turned a blind eye. No one told the fitter that on this job goggles were really necessary.

It is bad management for those issuing work permits to cover themselves by asking for more protective clothing than is really necessary. They should ask only for what is necessary and then *insist* that it be worn.

(b) Two men were told to wear breathing apparatus while repairing a compressor which handled gas containing hydrogen sulphide. The compressor had been swept out but traces of gas might have been left in it. One of the men had difficulty handling a heavy valve close to the floor and removed his mask. He was overcome by gas—hydrogen sulphide or possibly nitrogen.

Again, it is easy to blame the man. But he had been asked to do a job which was difficult wearing breathing apparatus. The plant staff resisted the temptation to blame him—the easy way out. Instead they looked for suitable lifting aids (7).

Similar incidents are discussed in Section 3.2. Rather than blame workers who make mistakes or disobey instructions we should try to remove the opportunities for error by changing the work situation, that is, the design, or method of operation.

(c) Work permits asked for goggles to be worn. They were not always worn and, inevitably, someone was injured. This incident differs from (a) in that the goggles *were* necessary in this case.

Investigation showed that the foreman and manager knew that goggles were not always worn. But they turned a blind eye to avoid dispute and to avoid delaying the job. The workers knew this and said to themselves, "Wearing goggles cannot be important." The foreman and manager were therefore responsible for the inevitable injury. People doing routine tasks become careless. Foremen and managers cannot be expected to stand over them all the time. But they can make occasional checks to see that the correct precautions are taken. And they can comment when they see rules being flouted. A friendly word *before* an accident is better than punitive action afterwards.

1.4.3 JOBS NEAR PLANT BOUNDARIES

Before a permit to weld or carry out other hot work is issued, it is normal practice to make sure that there are no leaks of flammable gas or liquid nearby, or no abnormal conditions that make a leak likely. The meaning of "nearby" depends on the nature of the material that might leak, the slope of the ground and so on. For highly flammable liquids, 15 m is often used.

Fires have occurred because a leak in one unit was set alight by welding in the unit next door. Before welding or other hot work is permitted within 15 m, say, of a unit boundary, the foreman of the unit next door should countersign it.

Similar hazards arise when a pipeline belonging to one unit passes through another unit.

Suppose a pipeline belonging to area A passes through area B and that this pipeline has to be broken in area B (Figure 1.9).

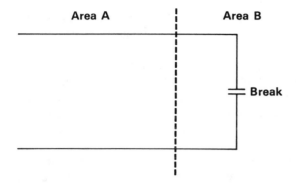

Figure 1.9. Who should authorize the pipeline break?

The person doing the job is exposed to two distinct hazards: those due to the contents of the pipeline: these are understood by Area A foreman, and those due to work going on in Area B: these are understood by Area B foreman.

If the work permit for the pipeline is issued by Area A foreman then Area B foreman should countersign it. If it is issued by B then A should countersign it. The system should be covered by local instructions and clearly understood.

An incident occurred because Area A foreman issued a permit for work to be done on a flow transmitter in a pipeline in Area B. Area B foreman issued a permit for grinding in area B. He checked that no flammable gas was present and had the drains covered. He did not know about the work on the flowmeter. A spark set fire to a drain line on the flowmeter, which had been left open.

What would happen in your plant?

1.4.4 MAINTENANCE WORK OVER WATER

A welder was constructing a new pipeline in a pipe trench, while 20 m away a slip-plate was being removed from another pipe which had contained light oil. Although the pipe had been blown with nitrogen, it was realized that a small amount of the oil would probably be spilled when the joint was broken. But it was believed that the vapor would not spread to the welders. Unfortunately, the pipe trench was flooded after heavy rain and the oil spread across the water surface and was ignited by the welder's torch. One of the men working on the slip-plate was badly burned and later died.

The first lesson from the incident is that *welding should not be allowed over large pools of water.* Spillages some distance away might be ignited.

In 1970, 35 tons of gasoline were spilled on the Manchester Ship Canal, England; 1 km away, 2½ hours later, the gasoline caught fire, killing six men (8).

The second lesson is that *when large joints have to be broken regularly, a proper means of draining the line should be provided.* The contents should not be allowed to spill onto the ground when the joint is broken.

Why was a permit issued to remove a slip-plate 20 m away from a welding job? Although vapor should not normally spread this far, the two jobs were rather close together.

The foremen who issued the two permits were primarily responsible for operating a unit some distance away. As they were busy with the running plant, they did not visit the pipe trench as often as they might. Had they visited it immediately before allowing the de-slip-plating job to start they would have realized that the two jobs were close together. They might have realized that oil would spread across the water in the trench.

After the incident, special day foremen were appointed to supervise construction jobs and interface with the construction teams. The construction teams like this system because they deal with only one process foreman instead of four shift foremen.

For another incident involving a construction team, see Section 5.4.2 (b).

1.4.5 MISUNDERSTANDINGS

Many incidents have occurred because of misunderstandings over the meanings of words and phrases. The following are two typical incidents.

(a) A permit was issued to remove a pump for overhaul. The pump was defused, removed and the open ends blanked. Next morning, the maintenance foreman signed the permit to show that the job— removing the pump—was complete. The morning shift lead operator glanced at the permit. Seeing that the job was complete, he asked the electrician to replace the fuses. The electrician replaced them and signed the permit to show that he had done so. By this time the afternoon shift lead operator had come on duty. He went out to check the pump and found that it was not there.

The job on the permit was to remove the pump for overhaul. Permits are sometimes issued to remove a pump, overhaul it, and replace it. But in this case the permit was just for removal (see Section 1.1.2). When the maintenance foreman signed the permit to show that the job was complete, he meant that the job of *removal* was complete. The lead operator, however, did not read the permit thoroughly. He assumed that the *overhaul* was complete.

The main message is clear: Read permits carefully; don't just glance at them.

When a maintenance worker signs a permit to show that the job is complete, he means he has completed the job *he thought he had to do*. This may not be the same as the job he was expected to do. The job should therefore always be inspected by the process team to make sure that the one completed is the one wanted.

When handing over or handing back a permit, the maintenance and process people should speak to each other. It is not good practice to leave a permit on the table for someone to sign when he comes in.

(b) When a work permit is issued to excavate the ground it is normal practice for an electrician to certify that there are no buried cables. What, however, is an "excavation"? A contractor asked for and received a work permit to "level and scrape the ground." No excavation was requested, so the process foreman did not consult the electricians. The contractor used a mechanical shovel, removed several feet from the ground and cut through a live electric cable. The word "excavation" needs careful definition.

1.4.6 A PERMIT TO WORK DANGEROUSLY?

A permit system is necessary for the safe conduct of maintenance operations. But issuing a permit in itself does not make the job safe. It merely provides an opportunity to check what has been done to make the equipment safe, to review the precautions necessary and to inform those who will have to carry out the job. That it is necessary to say this is shown by the following quotation from an official report.

> ". . . they found themselves in difficulty with the adjustment of some scrapers on heavy rollers. The firm's solution was to issue a work permit, but it was in fact a permit to live dangerously rather than a permit to work in safety. It permitted the fitter to work on the moving machinery with the guards removed. A second permit was issued to the first-aid man to enable him to stand close to the jaws of death ready to extricate, or die in the attempt to extricate, the poor fitter after he was dragged into the machinery. In fact, there was a simple solution. It was quite possible to extend the adjustment controls outside the guard so that the machinery could be adjusted, while still in motion, from a place of safety (9)."

1.5 QUALITY OF MAINTENANCE

Many accidents have occurred because maintenance work was not carried out in accordance with the (often unwritten) rules of good engineering practice, as the following incidents show.

1.5.1 THE RIGHT AND WRONG WAYS TO BREAK A JOINT

One of the causes of the fire described in Section 1.1.1 was the fact that the joint was broken incorrectly. The fitter removed all the nuts from the pump cover and then used a wedge to release the cover which was held tightly on the studs. It came off suddenly, followed by a stream of hot oil.

The correct way to break a joint is to slacken the nuts furthest away from you and then spring the joint faces apart, using a wedge if necessary. If any liquid or gas is present under pressure, then the pressure can be allowed to blow off slowly or the joint can be retightened.

It is not only flammable oils that cause accidents. In another incident two workers were badly scalded when removing the cover from a large valve on a hot water line, although the gauge pressure was only 9 inches of water (0.33 psi or 0.023 bar). They removed all the nuts, attached the cover to a chain block and tried to lift it. To release the cover they tried to rock it. The cover suddenly released itself and hot water flowed out onto the workers' legs.

1.5.2 USE OF EXCESSIVE FORCE

A joint on an 8-inch line containing a hot solvent had to be remade. The two sides were 3/4-inch out of line. There was a crane in the plant at the time so it was decided to use it to lift one of the lines slightly. The lifting strap pulled against a 3/4-inch branch and broke it off (Figure 1.10).

It was not a good idea to use a crane for a job like this on a line full of process material.

Fortunately the leaking vapor did not ignite, although nearby water was being pumped out of an excavation. At one time a diesel pump would have been used, but their use had been banned only a few months before the incident.

1.5.3 IGNORANCE OF MATERIAL STRENGTH

(a) When a plant came back on line after a long shutdown, some of the flanges had been secured with stud bolts and nuts instead of ordinary bolts and nuts. And some of the stud bolts were located so that

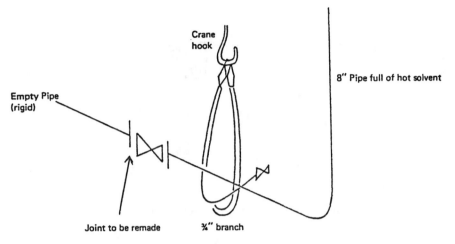

Crane
hook

8″ Pipe full of hot solvent

Empty Pipe
(rigid)

Joint to be remade   ¾″ branch

Figure 1.10. A branch broke when a crane was used to move a live line.

there was more protruding on one side than on the other. On some flanges, one of the nuts was secured by only two or three threads (Figure 1.11).

Nobody knows why this had been done. Probably one nut was tighter than the other and, in attempting to tighten this nut, the whole stud was screwed through the second nut. Whatever the reason, it produced a dangerous situation because the pressure on different parts of the flange was not the same.

In addition, stud bolts should not be indiscriminately mixed with ordinary bolts or used in their place. They are often made of different grades of steel and produce a different tension.

In the plant concerned, for the eight-bolt joints the bolts were changed one bolt at a time. Four-bolt joints were secured with clamps until the next shutdown.

(b) There was a leak on a large fuel-gas system operating at gasholder pressure. To avoid a shutdown, a wooden box was built around the leak and filled with concrete. It was intended as a temporary job but was so successful that it lasted for many years.

On other occasions leaks have been successfully boxed in or encased in concrete. But the operation can only be done at low pressures and expert advice is needed, as shown by the following incident.

There was a bad steam leak from the bonnet gasket of a 3-inch steam valve at a gauge pressure of 300 psi (20 bar). An attempt to

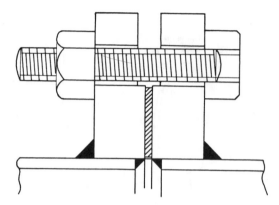

Figure 1.11. Nuts fitted incorrectly to studs.

clamp the bonnet was unsuccessful so the shift crew decided to encase the valve in a box. They made one 36 inches long, 24 inches wide and 14 inches deep out of ¼-inch steel plate. Plate of this thickness is strong but the shape of the box was unsuitable for pressure and could hardly have held a gauge pressure of more than 50 psi (3 bar), even if the welds had been full penetration, which they were not.

The box was fitted with a vent and valve. When the valve was closed the box started to swell and the valve was quickly opened.

A piece of 2-inch by 2-inch angle iron was then welded around the box to strengthen it. The vent valve was closed. A few minutes later the box exploded. Fortunately the mechanic—if he deserves the title—had moved away.

This did not happen in a back-street firm, but in a major international company.

These incidents show the need for continual vigilance. We cannot assume that because we employ qualified craftsmen and graduate engineers they will never carry out repairs in a foolish or unsafe manner.

1.5.4 FAILURE TO UNDERSTAND HOW THINGS WORK OR HOW THEY ARE CONSTRUCTED

(a) Several spillages have occurred from power operated valves while the actuators were being removed because the bolts holding the valve bonnets in position were removed in error. Figures 1.12 and 1.13 show how two such incidents occurred. The second system is

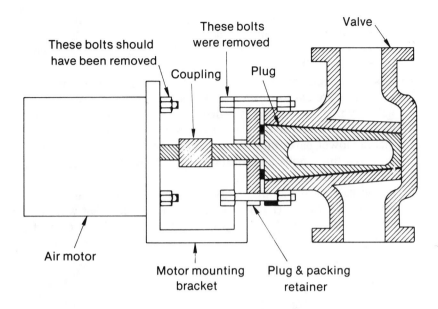

Figure 1.12. Wrong nuts undone to remove valve actuator.

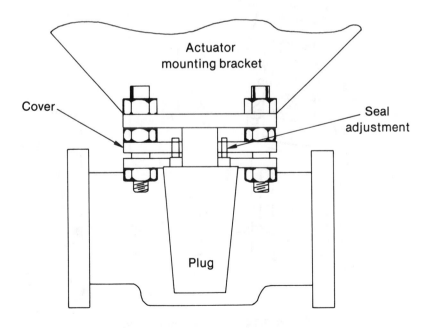

Figure 1.13. Wrong nuts undone to remove valve actuator.

particularly vulnerable because in trying to unscrew the nuts that hold the actuator mounting bracket in place, the stud may unscrew out of the lower nuts. This incident could be classified as due to poor design (10).

(b) A similar accident occurred on a common type of ball valve. Two workers were asked to fit a drain line below the valve. There was not very much room. So they decided to remove what they thought was a distance piece or adaptor below the valve but which was in fact the lower part of the valve body (Figure 1.14). When they had removed three bolts and loosened the fourth, it got dark and they left the job to the next day.

The valve was the drain valve on a small tank containing lique- fied petroleum gas (LPG). The five tons of LPG that were in the tank escaped over 2–3 hours but fortunately did not catch fire.

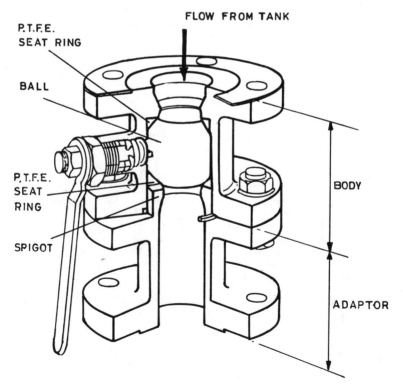

Figure 1.14. Valve dismantled in error.

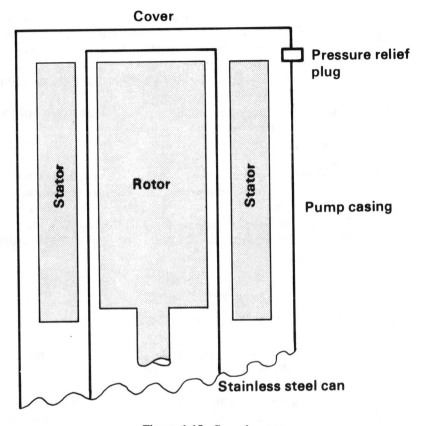

Figure 1.15. Canned pumps.

However 2,000 people who lived near the plant were evacuated from their homes (11).

(c) In canned pumps the moving part of the electric motor—the rotor—is immersed in the process liquid: there is no gland and gland leaks cannot occur.

The fixed part of the electric motor—the stator—is not immersed in the process liquid and is separated from the rotor by a stainless steel can (Figure 1.15).

If there is a hole in the can, process liquid can get into the stator compartment. A pressure relief plug is therefore fitted to the compartment and should be used before the compartment is opened for work on the stator. Warning plates, reminding us to do this, are often fitted to the pumps.

The stator compartment of a pump was opened up without the pressure relief plug being used. There was a hole in the can. This had caused a pressure buildup in the stator compartment. When the cover was unbolted it was blown off and hit a scaffold pole 2 m above. On the way up, it hit a man on the knee and the escaping process vapor caused eye irritation.

Persons working on the pump did not know the purpose of the plug and the warning notice was missing.

For a more detailed diagram and description of a canned pump, see Reference 12.

(d) On several occasions fitters have removed thermowells without realizing that this would result in a leak. They did not realize that the thermowell—the pocket into which a thermocouple or other temperature measuring device sits—is in direct contact with the process fluid. A serious fire which started this way is described in Reference 13.

1.5.5 TREATING THE SYMPTOMS INSTEAD OF THE DISEASE

The following incidents show what can happen if we go on repairing faults but never ask why so many faults occur.

(a) A cylinder lining on a high-pressure compressor was changed 27 times in nine years. On 11 occasions it was found to be cracked and on the other 16 occasions it showed signs of wear. No one asked why it had to be changed so often, they just went on changing it. Finally, a bit of the lining got caught between the piston and the cylinder head and split the cylinder.

(b) While a man was unbolting some 3/4-inch bolts one of them sheared. The sudden jerk caused a back strain and absence from work.

During the investigation of the accident, seven bolts were found nearby which had been similarly sheared on previous occasions. It was clear that the bolts sheared frequently. If, instead of simply replacing them and carrying on, the workers had reported the failures, then a more suitable bolt material could have been found.

Why did they not report the failures? If they had reported them would anything have been done?

The accident would not have occurred if the foreman or the engineer, on their plant tours, had noticed the broken bolts and asked why there were so many.

1.5.6 FLAMEPROOF ELECTRICAL EQUIPMENT

On many occasions, detailed inspections of flameproof electrical equipment have shown that many items were faulty. For example, at one plant a first look around indicated that nothing much was wrong. A more thorough inspection, paying particular attention to equipment not readily accessible, and which could be examined only from a ladder, showed that out of 121 items examined, 33 needed repair. The faults included missing and loose screws, gaps too large, broken glasses and incorrect glands. Not all the faults would have made the equipment a source of ignition, but many would have done so.

Why were there so many faults? Before this inspection there had been no regular inspections. Many electricians did not understand why flameproof equipment was used and what would happen if it was badly maintained. Spare screws and screwdrivers of the special types used were not in stock, so there was no way of replacing those lost.

Regular inspections were set up. Electricians were trained in the reasons why flameproof equipment is used and spares were stocked. In addition, it was found that in many cases flameproof equipment was not really necessary. Zone 2 equipment—cheaper to buy and easier to maintain—could be used instead.

REFERENCES

1. T. A. Kletz in H. H. Fawcett and W. S. Wood, (editors), *Safety in Chemical Operations,*" 2nd edition, Wiley, 1982, chapter 26.
2. F. P. Lees, *Loss Prevention in the Process Industries,* Butterworths, 1980, chapter 21.
3. T. A. Kletz, *Proceedings of the Second International Symposium on Loss Prevention and Safety Promotion in the Process Industries,* Heidelberg, Sept. 1977, Dechema, Frankfurt, p. 1.
4. *Annual Report of the Chief Inspector of Factories for 1974,* Her Majesty's Stationery Office, London, 1975, p. 19.
5. *The Bulletin, The Journal of the Society for Petroleum Acts Administration,* Oct. 1970, p. 68.
6. *Chemical Safety Summary,* Chemical Industries Association, London, July–Sept 1980, p. 15.
7. *Petroleum Review,* April 1982, p. 34.
8. *Annual Report of Her Majesty's Inspectors of Explosives for 1970,* Her Majesty's Stationery Office, London, 1971, p. 19.

9. *Health and Safety—Manufacturing and Service Industries 1979,* Her Majesty's Stationery Office, London, 1981, p. 62.

10. T. A. Kletz, *Hydrocarbon Processing,* Vol. 61, No.3, March 1982, p. 207.

11. *Leakage of Propane at Whitefriars Glass Limited, Wealdstone, Middlesex, 20 November 1980,* Health and Safety Executive, London, 1981.

12. G. R. Webster, *The Chemical Engineer,* Feb. 1979, p. 91.

13. *Petroleum Review,* Oct. 1981, p. 21.

Chapter 2

Modifications

Many accidents have occurred because changes were made in plants and these changes had unforeseen side effects. In this chapter a number of such incidents are described. How to prevent similar changes in the future is discussed. Some of the incidents are taken from References 1 and 2 where others are described.

2.1 STARTUP MODIFICATIONS

Startup is a time when many modifications may have to be made. It is always a time of intense pressure. It is therefore not surprising that some modifications introduced during startup have had serious unforeseen consequences.

At one plant, a repeat relief and blowdown review was carried out one year after startup. The startup team had been well aware of the need to look for the consequences of modifications and had tried to do so as they were made. Nevertheless, the repeat relief and blowdown review brought to light 12 instances in which the assumptions of the original review were no longer true, and additional or larger relief valves, or changes in the position of a relief valve, were necessary. Figure 2.1 shows some examples.

The line diagrams had been kept up-to-date despite the pressures on the plant staff during startup. This made it easier to repeat the relief and blowdown review. The plant staff was so impressed by the results that it decided to have another look at the relief and blowdown after another year.

A late modification which had unforeseen results is described in Section 5.5.2 (c).

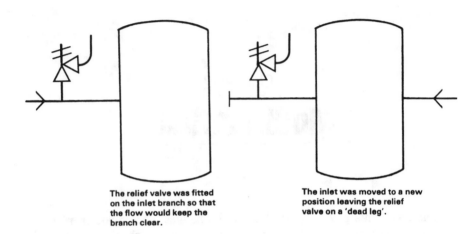

The relief valve was fitted on the inlet branch so that the flow would keep the branch clear.

The inlet was moved to a new position leaving the relief valve on a 'dead leg'.

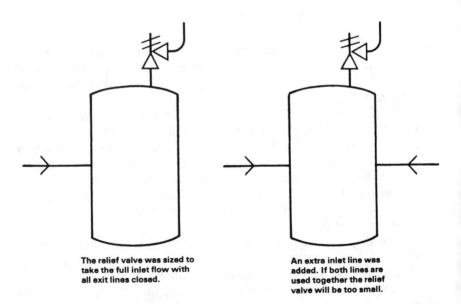

The relief valve was sized to take the full inlet flow with all exit lines closed.

An extra inlet line was added. If both lines are used together the relief valve will be too small.

Figure 2.1. Some of the modifications made to the relief system in a plant during its first year on-line.

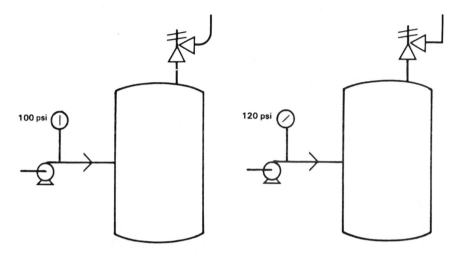

The vessel was designed to withstand the maximum pressure the pump could deliver. The relief valve was not designed to take the maximum flow from the pump.

The pump actually proved capable of producing 20 psi more than design. If the exit from the vessel is isolated when the pump is running the vessel will be overpressured.

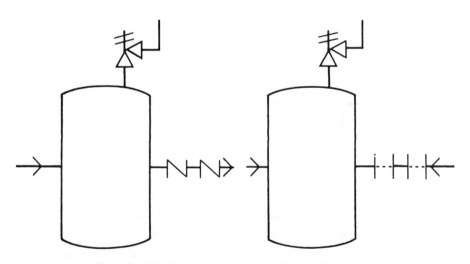

The relief valve was sized on the assumption that two non-return valves in series would prevent back-flow into the vessel.

Both non-return valves corroded, allowing back-flow to take place.

Figure 2.1. Continued.

(Continued on next page.)

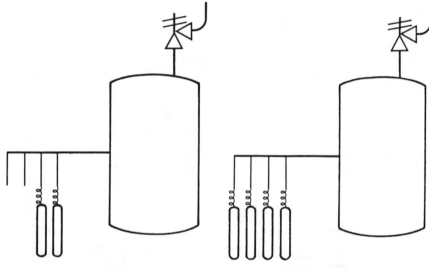

The relief valve was sized on the assumption that only two gas cylinders would be used at a time, though connections were provided for four cylinders.

Inevitably, four cylinders were connected up and sometimes used.

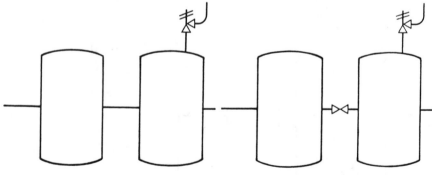

A single relief valve was designed to protect two vessels which were connected together by a line without any valve or other restriction between them.

An extra isolation valve was fitted between the two vessels, thus making it possible to isolate the first vessel from its relief valve.

In another similar case, chokes occurred in the line between the two vessels.

Figure 2.1. Continued.

2.2 MINOR MODIFICATIONS

This term is used to describe modifications so inexpensive that they either do not require formal financial sanction or the sanction is easily obtained. They therefore may not receive the same detailed consideration as a more expensive modification.

(a) A modification so simple that it required only a work permit resulted in the end blowing off a tank and fatal injuries to two men working in the area.

The tank was used for storing a liquid product that melts at 97°C. It was therefore heated by a steam coil using steam at a gauge pressure of 100 psi (7 bar). At the time of the incident, the tank was almost empty and was being prepared to receive some product. The inlet line was being blown with compressed air to prove that it was clear—the normal procedure. The air was not getting through and the operator suspected a choke in the pipeline.

In fact, the vent on the tank was choked. The gauge air pressure (75 psi or 5 bar) was sufficient to burst the tank (design gauge pressure 5 psi or 0.3 bar).

Originally the tank had a 6-inch-diameter vent. But at some time this was blanked off and a 3-inch-diameter dip branch was used instead as the vent.

There were several other things wrong. The vent was not heated; its location made it difficult to inspect; most important of all, neither manager, supervisors nor operators recognized that if the vent choked, the air pressure was sufficient to burst the tank. Nevertheless, if the 6-inch vent had not been blanked the incident would not have occurred. See also Section 12.1.

(b) A reactor was fitted with a bypass (Figure 2.2). The remotely-operated valves A, B, and C were interlocked so that C had to be open before A or B could be closed. It was found that the valves leaked, so hand-operated isolation valves (a, b and c) were installed in series with them (Figure 2.3).

After closing A and B, the operators were instructed to go outside and close the corresponding hand valves a and b. This destroyed the interlocking. One day an operator could not get A and B to close. He had forgotten to open C. He decided that A and B were faulty and closed a and b. Flow stopped. The tubes in the furnace were overheated. One of them burst and the lives of the rest were shortened.

(c) Other minor modifications which have had serious effects on plant safety are:

1. Removing a restriction plate which limits the flow into a vessel and which has been taken into account when sizing the vessel's relief valve.
2. Fitting a larger trim into a control valve when the size of the trim limits the flow into a vessel and has been taken into account when sizing the vessel's relief valve.
3. Fitting a substandard drain valve. See Section 8.2 (a).
4. Replacing a metal duct by a hose. See Section 15.3.

2.3 MODIFICATIONS MADE DURING MAINTENANCE

Even when systems for controlling modifications have been set up, modifications often slip in unchecked during maintenance. (Someone decides, for what he thinks is a good reason, to put the plant back slightly differently.)

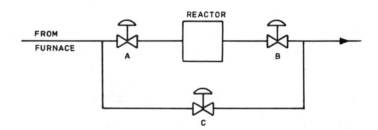

Figure 2.2. Original reactor bypass.

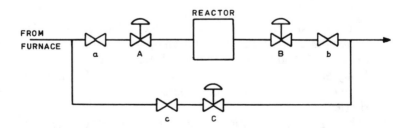

Figure 2.3. Modified reactor bypass.

Thirty years ago a special network of air lines was installed for use with breathing apparatus only. A special branch was taken off the top of the compressed air main as it entered the Works (Figure 2.4).

For thirty years this system was used without any complaint. Then one day a man got a faceful of water while wearing a face mask inside a vessel. Fortunately he was able to signal to the stand-by man that something was wrong and he was rescued before he suffered any harm.

On investigation, it was found that the compressed air main had been renewed and that the branch to the breathing apparatus network had been moved to the *bottom* of the main. When a slug of water got into the main, it all went into the catchpot which filled up more quickly than it could empty. Unfortunately, everyone had forgotten why the branch came off the top of the main and nobody realized that this was important.

A very similar incident occurred on a fuel-gas system. When a corroded main was renewed, a branch to a furnace was taken off the bottom

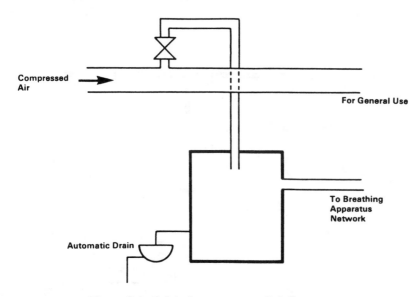

Figure 2.4. Original arrangement of air lines.

of the main instead of the top. A slug of liquid filled up the catchpot and extinguished the burners.

2.4 TEMPORARY MODIFICATIONS

The most famous of all temporary modifications is the temporary pipe installed in the Nypro Factory at Flixborough, UK, in 1974. It failed two months later, causing the release of about 50 tons of hot cyclohexane. The cyclohexane mixed with the air and exploded, killing 28 people and destroying the plant (3).

At the Flixborough plant there were six reactors in series. Each reactor was slightly lower than the one before so that the liquid in them flowed by gravity from No.1 down to No.6 through short 28-inch-diameter connecting pipes (Figure 2.5). To allow for expansion, each 28-inch pipe contained a bellows.

One of the reactors developed a crack and had to be removed. (The crack was the result of a process modification; see Section 2.6 (b).) It was replaced by a temporary 20-inch pipe, which had two bends in it, to allow for the difference in height. The existing bellows were left in position at each end of the temporary pipe (Figure 2.5).

The design of the pipe and support left much to be desired. The pipe was not properly supported; it merely rested on scaffolding. Because there was a bellows at each end it was free to rotate or "squirm" and did so when the pressure rose a little above the normal level. This caused the bellows to fail.

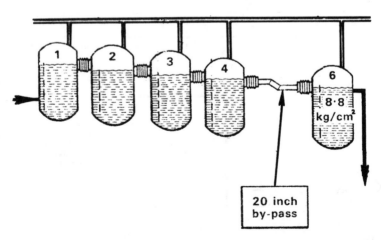

Figure 2.5. Arrangement of reactors and temporary pipe at Flixborough.

There was no professionally qualified engineer in the plant at the time the temporary pipe was built. The men who designed and built it (design is hardly the word because the only drawing was a full-scale sketch in chalk on the workshop floor) did not know how to design large pipes which are required to operate at high temperatures (150°C) and gauge pressures (150 psi or 10 bar). Very few engineers have the specialized knowledge to design highly-stressed piping. But in addition, the engineers at Flixborough did not know that design by experts was necessary. They did not know what they did not know (3).

For another temporary modification see Section 5.5.1.

2.5 SANCTIONED MODIFICATIONS

This term is used to describe modifications for which the money has to be authorized by a senior manager or a committee. They cannot, therefore, be done in a hurry. Justifications have to be written out and people persuaded. Although the systems are (or have been in the past) designed primarily to control cost rather than safety, they usually result in careful consideration of the proposal by technical personnel. Unforeseen consequences may come to mind, though not always. Sometimes sanction is obtained before detailed design has been carried out and the design may then escape detailed considerations. Nevertheless, it is harder to find examples of serious incidents caused by sanctioned modifications. The following might almost rank as a startup modification. Though the change was agreed over a year before startup, it occurred after the initial design had been studied and approved.

(a) A low-pressure refrigerated ethylene tank was provided with a relief valve set at a gauge pressure of about 1.5 psi (0.1 bar) and discharging to a vent stack. After the design had been completed, it was realized that cold gas coming out of the stack would, when the wind speed was low, drift down to ground level where it might be ignited. The stack was too low to be used as a flare stack—the radiation at ground level would be too high—and was not strong enough to be extended. What could be done?

Someone suggested putting steam up the stack to disperse the cold vapor. This seemed a good idea and the suggestion was adopted (Figure 2.6).

As the cold vapor flowed up the stack it met condensate flowing down. The condensate froze and completely blocked the 8-inch-diameter stack. The tank was over-pressured and ruptured. Fortunately the rupture was a small one and the escaping ethylene did not ignite. It was dispersed with steam while the tank was emptied.

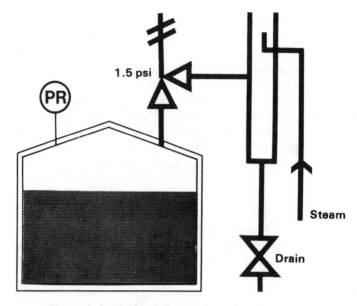

Figure 2.6. Liquid ethylene tank venting arrangements.

Should the design team have foreseen that the condensate might freeze?

After the tank was repaired, the vent stack was replaced by a flare stack.

See also Sections 3.3.1 (b) and 6.2 (b).

(b) A new loading gantry was built for filling tank trucks with lique-fied petroleum gas. The ground was sloped so that any spillages would run away from the tanker and would not heat it if they caught fire. As a result of this change in design, it was found that the level indicator would not read correctly when the tank truck was located on sloping ground. The design had to be modified again so that the wheels stood on level ground but the ground in between and around them was sloped.

2.6 PROCESS MODIFICATIONS

So far we have discussed modifications to the plant equipment. Acci-dents can also occur because changes to process materials or conditions had unforeseen results, as the following cases show:

(a) A slight change in raw material quality caused a big loss in production.

A hydrogenation reactor developed a pressure drop. Various causes were considered—catalyst quality, size, distribution and activation; reactant quality, distribution and degradation—before the true cause was found.

The hydrogen came from another plant and was passed through a charcoal filter to remove traces of oil before it left the supplying plant. Changes of charcoal were infrequent, and the initial stock lasted several years. Reordering resulted in a finer charcoal being supplied and charged without the significance of the change being recognized. Over a long period, the new charcoal passed through its support into the line to the other plant. Small amounts of the charcoal partially clogged the 3/8-inch distribution holes in the catalyst retaining plate (Figure 2.7).

The cause of the pressure drop was difficult to find because it was due to a change in another plant.

(b) At one time it was common to pour water over equipment which was too hot, or which was leaking fumes. The water was taken from the nearest convenient supply. At Flixborough, there was a leak of cyclohexane vapor from the stirrer gland on one of the reactors. To condense the leaking vapor, water was poured over the top

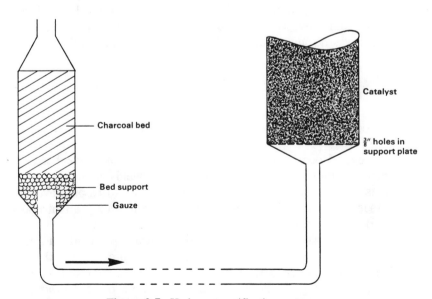

Figure 2.7. Hydrogen purification system.

of the reactor. Plant cooling water was used because it was conveniently available.

Unfortunately, the water contained nitrates which caused stress corrosion cracking of the mild steel reactor. The reactor was removed for repair and the temporary pipe which replaced it later failed and caused the explosion. See Section 2.4.

Nitrate-induced cracking is well known to metallurgists but was not well known to other engineers at the time. Before you poured water over equipment—emergencies apart—would you ask what the water contained and what its effect would be on the equipment?

Pouring water over equipment is a change outside normal operating practice. It should therefore be treated as a modification.

For a description of nitrate cracking of mild steel, see Reference 4.

(c) The following incident shows how difficult it is to foresee all the results of a change and how effects can be produced a long way downstream of the place where the change is made.

Some radioactive bromine (half-life 36 hours), in the form of ammonium bromide, was put into a brine stream as a radioactive tracer. At another plant 30 km away the brine stream was electrolyzed to produce chlorine. Radioactive bromine entered the chlorine stream and subsequently concentrated in the base of a distillation column which removed heavy ends. This column was fitted with a radioactive-level controller. The radioactive bromine affected the level controller which registered a low level and closed the bottom valve on the column. The column became flooded. There was no injury, but production was interrupted.

2.7 NEW TOOLS

The introduction of new tools can have unforeseen side effects.

(a) On several occasions radioactive level indicators have been affected by radiography being carried out on welds up to 70 m away.

(b) This incident did not occur in the process industries but nevertheless is a good example of the way in which a new tool can introduce unforeseen hazards.

A natural gas company employed a contractor to install a 2-inch plastic natural gas main to operate at a gauge pressure of 60 psi (4 bar) along a street. The contractor used a pneumatic boring technique. In doing so, he bored right through a 6-inch sewer pipe serving one of the houses in the street.

The occupant of the house, finding that his sewer was obstructed, engaged another contractor to clear it. The contractor used an auger and ruptured the plastic gas pipe. Within three minutes the natural gas had traveled 12 m up the sewer pipe into the house and exploded. Two people were killed and four injured. The house was destroyed and the houses on either side were damaged.

After the explosion it was found that the gas main had passed through a number of other sewer pipes (5).

2.8 ORGANIZATIONAL CHANGES

These can also have unforeseen side effects as shown by the following incidents.

(a) A plant used sulphuric acid and caustic soda in small quantities so they were supplied in similar plastic containers called polycrates (Figure 2.8). While an operator was on his day off someone decided it would be more convenient to have a polycrate of acid and a polycrate of alkali on each side (Figure 2.9).

When the operator came back, no one told him about the change. Without checking the labels he poured some excess acid into a caustic crate. There was a violent reaction and the operator was sprayed in the face. Fortunately he was wearing goggles.

We should tell people about changes made while they were away.

In addition, if incompatible chemicals are handled at the same plant then, whenever possible, the containers should differ in size,

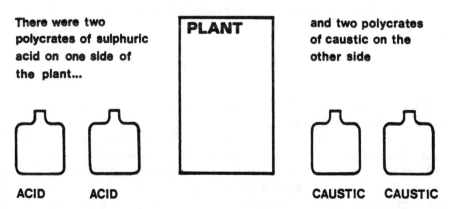

There were two polycrates of sulphuric acid on one side of the plant...

PLANT

and two polycrates of caustic on the other side

ACID   ACID   CAUSTIC   CAUSTIC

Figure 2.8. Original layout of acid and caustic containers.

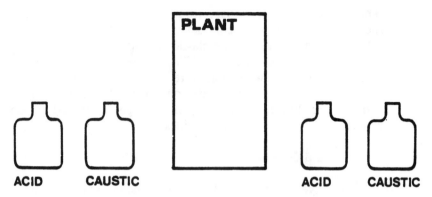

Figure 2.9. Modified layout of acid and caustic containers.

shape and/or color and the labels should be large and easily seen from eye level.

(b) The staff of a plant decided to exhibit work permits so that they can be more readily seen by workers on the job—a good idea.

The permits were usually put in plastic bags and tied to the equipment. But sometimes they were rolled up and inserted into the open ends of scaffold poles.

One day a man put a permit into the open end of a pipe. He probably thought that it was a scaffold pole or defunct pipe. Unfortunately it was the air bleed into a vacuum system. The air rate was controlled by a motor valve. The permit got sucked into the valve and blocked it. The vacuum could not be broken, product was sucked into the vacuum system and the plant had to be shut down for cleaning for two days.

2.9 GRADUAL CHANGES

These are the most difficult to control. Often we do not realize that a change is taking place until it is too late. For example, over the years, steam consumption at a plant had gradually fallen. Flows through the mains became too low to prevent condensate accumulating. On one of the mains an inaccessible steam trap had been isolated and the other main had settled slightly. Neither of these mattered when the steam flow was large, but it gradually fell. Condensate accumulated and finally water hammer fractured the mains.

2.10 CONTROL OF MODIFICATIONS

How can we prevent modifications producing unforeseen and undesirable side effects? References 1 and 2 proposed a three-pronged approach.

(1) Before any modification, however inexpensive, temporary or permanent, is made to a plant or process or to a safety procedure it should be authorized in writing by a manager *and* an engineer— that is, by professionally qualified staff, usually the first level of professionally qualified staff.

(2) The managers and engineers who authorize modifications cannot be expected to stare at a drawing and hope that the consequences will show up. They must be provided with an aid such as a list of questions to be answered. Such an aid is shown in References 1 and 2.

(3) It is not sufficient to issue instructions about (1) and the aid described in (2). We must convince all concerned, particularly foremen, that they should not carry out unauthorized modifications. This can be done by discussing typical incidents such as those described here or those illustrated in the Institution of Chemical Engineers (UK) *Hazard Workshop Module* No. 002.

REFERENCES

1. T. A. Kletz, *Chemical Engineering Progress,* Vol.72, No.11, Nov. 1976, p.48.
2. F. P. Lees, *Loss Prevention in the Process Industries,* Butterworths, 1980, Vol.2, Chapter 21.
3. *The Flixborough Cyclohexane Disaster,* Her Majesty's Stationery Office, London, 1975.
4. *Guide Notes on the Safe use of Stainless Steel,* Institution of Chemical Engineers, Rugby, U.K., 1978.
5. A note issued by the U.S. National Transportation Safety Board on Nov. 12, 1976.

Chapter 3

Accidents Caused by Human Error

3.1 INTRODUCTION

This Chapter describes accidents due to those aberrations that even well-trained and well-motivated persons make from time to time. For example, they forget to close a valve or close the wrong valve. They know what they should do, want to do it, and are physically and mentally capable of doing it. But they forget to do it. Exhortation, punishment, or further training will have no effect. We must either accept an occasional mistake or change the work situation so as to remove the opportunities for error or make errors less likely.

Many of the errors occur, not in spite of the fact that the man is well-trained, but *because* he is well-trained. Routine operations are relegated to the lower levels of the brain and are not continuously monitored by the conscious mind. We would never get through the day if everything we did required our full attention. When the normal pattern or program of actions is interrupted for any reason, errors are likely to occur. These slips are very similar to those we make in everyday life. Their psychology has been described by Reason and Mycielska (1).

We then describe some accidents which occurred because employees were not adequately trained. Sometimes they lacked basic knowledge; sometimes they lacked sophisticated skills.

Errors also occur because people deliberately decide not to carry out instructions which they consider unnecessary. In particular, they do not wear protective clothing or take other precautions detailed on a work permit. These errors were discussed in Section 1.4.2, where three points were made:

(a) Are the instructions really necessary? If not, they should be canceled.
(b) Are the instructions practicable? If not, a better method should be devised.
(c) If the answers to (a) and (b) are yes, then are regular checks made to see that the instructions are followed?

3.2 ACCIDENTS CAUSED BY SIMPLE SLIPS. TO PREVENT THEM WE SHOULD CHANGE THE PLANT DESIGN OR METHOD OF WORKING

3.2.1 "THERE IS NOTHING WRONG WITH THE DESIGN. THE EQUIPMENT WASN'T ASSEMBLED CORRECTLY."

How often has this been said by the designer after a piece of equipment has failed? The designer is usually correct, but whenever possible we should use designs that are impossible (or difficult) to assemble wrongly, or are unlikely to fail if they are assembled wrongly. For example:

(a) In some compressors it is possible to interchange suction and delivery valves. Damage and leaks have developed as a result. Valves should be designed so that they cannot be interchanged.
(b) With many types of screwed couplings and compression couplings it is easy to use the wrong ring. Accidents have occurred as a result. Flanged or welded pipes should therefore be used except for small-bore lines carrying nonhazardous materials.
(c) Loose-backing flanges require more care during joint making than fixed flanges. Fixed flanges are therefore preferred.
(d) Bellows should be installed with great care because unless specially designed, they cannot withstand any sideways thrust. With hazardous materials, it is therefore good practice to avoid the need for bellows by designing expansion bends into the pipework.
(e) A runaway reaction occurred in a polymerization reactor. A rupture disc failed to burst. It had been fitted on the wrong side of the vacuum support, thus raising its bursting pressure from a gauge pressure of 150 psi (10 bar) to about 400 psi (27 bar) (Figures 3.1A and 3.1B).
   The polymer escaped through some of the flanged joints, burying the reactor in a brown polymer that looked like toffee (molasses candy). The reactor was fitted with class 150 flanges. If these are overpressured, the bolts will stretch and the flanges will leak, thus preventing the vessel from bursting (provided the pressure

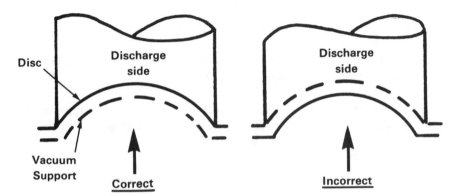

Figure 3.1A. Arrangements of rupture disc and vacuum support.

Figure 3.1B. Because a rupture disc was fitted to the wrong side of a vacuum support, the flanges leaked covering the reactor with "candy."

does not rise too rapidly). But this may not occur with flanges of a higher pressure rating.

The best way to prevent accidents such as this is to use rupture discs which are harder to assemble incorrectly and which can be checked for correct installation after assembly. It is possible to get discs permanently attached to their vacuum supports by the manufacturer and fitted with a projecting tag which carries the words "vent side" on one side. The tag also gives the pressure rating.

See Section 5.3, item 7, and Section 9.1.3.

3.2.2 WRONG VALVE OPENED

The pump feeding an oil stream to the tubes of a furnace failed. The operator closed the oil valve and intended to open a steam valve to purge the furnace tubes. He opened the wrong valve, there was no flow to the furnace, and the tubes were overheated and collapsed.

This incident is typical of those that would at one time have been blamed on human failing—the operator was at fault and there was nothing anyone else could do. In fact investigation showed that:

1. The access to the steam valve was poor and it was difficult to see which was the right valve.
2. There was no indication in the control room to show that there was no flow through the furnace coils.
3. There was no low-flow alarm or trip on the furnace.

3.2.3 WOULD YOU CLIMB OVER A PIPE OR WALK 90 m?

To repair a flowmeter a man had to walk six times from the orifice plate to the transmitter and back. To get from one to the other he had to walk 45 m, cross a 30-inch-diameter pipe by a footbridge and walk 45 m back—a total of 540 m for the whole job. Alternatively he could climb over the pipe.

While doing so he hurt his back.

Is it reasonable to expect a man to repeatedly walk 90 m to avoid climbing over a pipe?

3.2.4 AN ERROR WHILE TESTING AN ALARM

Two furnaces were each fitted with a temperature recorder controller and high temperature alarm. The two recorders were side by side on the instrument panel in the control room with the recorder for A furnace on the left (Figure 3.2).

An instrument mechanic was asked to test the alarm on A furnace. He put the controller on manual and then went behind the panel. The next step was to take the cover off a junction box, disconnect one of the leads, apply a gradually increasing potential from a potentiometer and note the reading at which the alarm sounded.

Behind the panel the junction boxes for A and B are in line with the recorders and therefore B is on the left (Figure 3.3).

The only label was very small and close to the floor so it was hardly readable.

The mechanic, who had done the job many times before, took the cover off B junction box and disconnected one of the leads. The effect was the same as if the thermocouple had burned out. The recorder registered a high temperature, the controller closed the fuel gas valve and the furnace tripped.

The two junction boxes should have been labeled A and B in large letters. Better still, the connections for the potentiometer should have been on the front of the panel.

3.2.5 POOR LAYOUT OF INSTRUCTIONS

A batch went wrong. Investigation showed that the operator had charged 104 kg of one constituent instead of 104 g (0.104 kg).

The instructions to the operator were set out as follows (the names of the ingredients being changed):

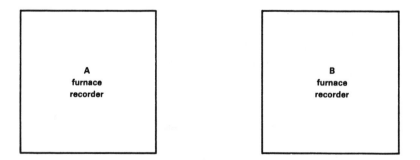

Figure 3.2. Layout of recorders on panel.

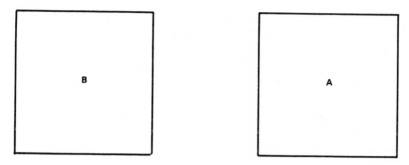

Figure 3.3. Layout of recorders behind panel.

Blending Ingredients	Quantity tons
Marmalade	3.75
Oxtail soup	0.250
Pepper	0.104 kg
Baked beans	0.020
Raspberry jam	0.006
TOTAL	4.026

With instructions like these it is very easy for the operator to get confused.

3.2.6 AN INACCURATE READING NOT NOTICED ON AN INSTRUMENT AT THIGH LEVEL

A reactor was being started up. It was filled with reaction mixture from another reactor which was already on line and the panel operator started to add fresh feed, gradually increasing the flow while he watched the temperature on a recorder conveniently situated at eye level. He intended to start a flow of cooling water to the reaction cooler as soon as the temperature started to rise—the usual method.

Unfortunately, there was a fault in the temperature recorder and although the temperature actually rose this was not indicated. Result: A runaway reaction.

The rise in temperature was, however, indicated on a six-point temperature recorder at a lower level on the panel, but the operator did not notice this (Figure 3.4).

An interesting feature of this incident was that no one blamed the operator. The manager said he would probably have made the same mistake because the check instrument was at a low level (about 1 m above the floor) and because a change in one temperature on a six-point recorder in that position is not obvious unless you are actually looking for it. It is not the sort of thing you notice out of the corner of your eye.

TRC Instrument

6 Point Temperature Recorder

Figure 3.4. Instruments below eye level may not be noticed.

3.2.7 CLOSING A VALVE IN ERROR CAUSED AN EXPLOSION

Figure 3.5 shows part of a plant in which there were five reactors in parallel. There were two gas-feed lines with cross connections between them. Oxygen was also fed to the reactors, but the oxygen lines are not shown. At the time of the incident only two reactors, Nos. 1 and 4, were on line.

The operator thought valve B was open so he shut valve A. This stopped the flow of gas to No. 1 reactor. The oxygen flow was controlled by a ratio controller, but it had a zero error and a small flow of oxygen continued.

When the operator realized his mistake and restored the gas flow, the reactor contained excess oxygen and an explosion occurred, not actually in the reactor but in the downstream waste heat boiler. Four men were killed.

Here we have a situation where simple error by an operator produced serious consequences. The explosion was not, however, the operator's fault, but the result of bad design and lack of protective equipment.

We would never knowingly tolerate a situation in which accidental operation of a valve resulted in the overpressuring of a vessel. We would install a relief valve. In the same way, accidental operation of a valve should not be allowed to result in explosion or runaway reaction.

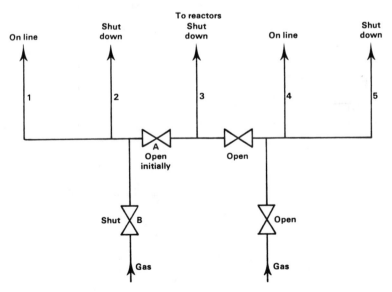

Figure 3.5. Accidental closing of a valve can cause an explosion.

3.2.8 AN EXPLOSION IN A BATCH REACTOR

Figure 3.6 shows a batch reaction system. A batch of glycerol was placed in the reactor and circulated through a heat exchanger which could act as both a heater and a cooler. Initially it was used as a heater and when the temperature reached 115°C, addition of ethylene oxide was started. The reaction was exothermic and the exchanger was now used as a cooler.

The ethylene oxide pump could not be started unless:

(a) The circulation pump was running.
(b) The temperature was above 115°C, as otherwise the ethylene oxide would not react.

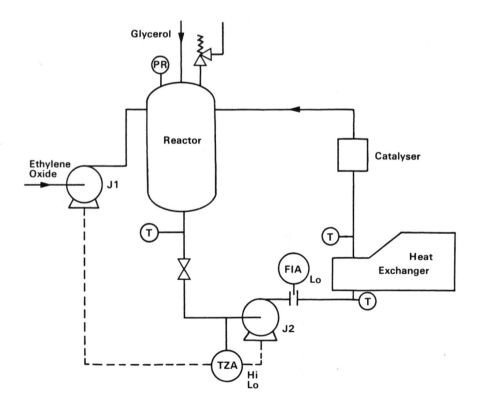

Figure 3.6. Arrangement of reactor circulating system.

(c) The temperature was below 125°C, as otherwise the reaction was too fast.

Despite these precautions an explosion occurred. One day, when ethylene oxide addition was started, the pressure in the reactor rose. This showed that the ethylene oxide was not reacting. The operator decided that perhaps the temperature point was reading low, or perhaps a bit more heat was required to start the reaction, so he adjusted the trip setting and allowed the indicated temperature to rise to 200°C. Still the pressure did not fall.

He then suspected that his theory might be wrong. Perhaps he had forgotten to open the valve at the base of the reactor? He found it shut and opened it. Three tons of unreacted ethylene oxide, together with the glycerol, passed through the heater and catalyzer and a violent uncontrolled reaction occurred. The reactor burst and the escaping gases exploded. Two men were injured. One, 160 m away, was hit by flying debris, and the other was blown off the top of a tanker.

Although the indicated temperature had risen, the temperature of the reactor's contents had not. Pump J2, running with a closed suction valve, got hot and the heat affected the temperature point, which was close to the pump.

Why did it happen?

1) The explosion was due to an operator forgetting to open a valve. It was not due to lack of knowledge, training, or instructions, but was another of those mistakes that even well-trained, well-motivated, capable people make from time to time.

2) If the operator had not opened the valve when he found it shut, the explosion could have been avoided. However, it is hard to blame him. His action was instinctive. What would you do if you found undone something you should have done some time ago?

3) The explosion was due to a failure to heed warning signs. The high pressure in the reactor was an early warning but the operator had another theory to explain it. He stuck to this theory until the evidence against it was overwhelming. This is known as a mind-set or tunnel vision.

   The other temperature points would have helped the operator to diagnose the trouble. But he did not look at them. He probably thought there was no point in doing so. All the temperature points were bound to read the same. The need for checking one reading by another should have been covered in the operator's training.

4) The explosion was due to a failure to directly measure the property we wish to know. The temperature point was not measuring the temperature in the reactor but the temperature near the pump. This

got hot because the pump was running with a closed suction valve. Similarly, the trip initiator on J2 showed that its motor was energized. It did not prove that there was a flow.

5) The explosion occurred because key instruments were not kept in working order. The flow indicator and low flow alarm (FIA) were out of order. They often were, and the operators had found that the plant could be operated without them. If there is no flow, they thought, J2 will have stopped and this will stop J1.

6) The operator should not have raised the trip setting, though doing so did not in itself cause the explosion. (However, he did try to use his intelligence and think why reaction was not occurring. Unfortunately, he was wrong.)

What should we do?

It is no use telling the operator to be more careful. We have to recognize that the possibility of a mistake—forgetting to open the valve—is inherent in the work situation. If we want to prevent a mistake, we must change the work situation. That is, change the design and/or the method of operation—the hardware and/or the software.

The original report blamed the operator for the explosion. But his failure to open the valve might have been foreseen.

1) The temperature should be measured in the reactor or as close to it as possible. We should always try to measure the property we wish to know directly, rather than another property from which the property we wish to know can be inferred.

   The designers assumed that the temperature near the pump would be the same as that in the reactor. It will not be if there is no circulation.

   The designers assumed that if the pump is energized, then liquid is circulating, but this not always the case.

2) Operators should not be allowed to change trip settings at will. Different temperatures are needed for different batches. But even so, the adjustment should be made only by someone who is given written permission to do so.

3) More effort might have been made to keep the flow indicator alarm in working order.

4) A high-pressure trip should be installed on the reactor.

5) Operators should be trained to "look before they leap" when they find valves wrongly set. See also Section 3.3.5 (a). Other accidents which occurred because operators did not carry out simple tasks are described in Sections 17.1 and 13.5.

3.3 ACCIDENTS WHICH COULD BE PREVENTED BY BETTER TRAINING

As we shall see, very often it is not a lack of sophisticated training that results in accidents but ignorance of the basic requirements of the job or the basic properties of the materials and equipment handled.

3.3.1 READINGS IGNORED

Many accidents have occurred because operators apparently thought their job was just to write down readings and not to respond to them.

(a) The temperature controller on the base of a distillation column went out of order at 5 a.m. and drew a straight line. This was not noticed. During the next seven hours the following readings were abnormal:
1. Six tray temperatures (one rose from 145°C to 255°C)
2. Level in base of still (low)
3. Level in reflux drum (high)
4. Take-off rate from reflux drum (high).

Most of these parameters were recorded on the panel. All were written down by the operator on the record sheet.

Finally, at 12 noon, the reflux drum overflowed and there was a spillage of flammable oil.

From 7 a.m. onward the operator was a trainee but a lead operator was present and the foreman visited the control room from time to time.

(b) Section 2.5 (a) described how a low-pressure liquefied ethylene storage tank split when the vent pipe became plugged with ice. For 11 hours before the split occurred the gauge pressure in the tank was reading 2 psi (0.13 bar). This pressure was above the set-point of the relief valve (a gauge pressure of 1.5 psi or 0.1 bar) and was the full-scale reading on the pressure gauge. The operators entered this reading on the record sheet but took no other action and did not even draw it to the attention of the foremen or managers when they visited the control room (2).

(c) The governor assembly and guard on a steam engine disintegrated with a loud bang, scattering bits over the floor. Fortunately no one was injured.

It was then found that the lubricating oil gauge pressure had been only 8 psi (0.5 bar) instead of 25 psi (1.7 bar) for at least "several months." In this case the pressure was not written down on the record sheet.

(d) The level measuring instrument and alarm on a feed tank were out of order, so the tank was hand-dipped every shift. When the plant was shut down the operators stopped dipping the tank.

The plant which supplied the feed was not shut down. It continued to supply feed into the feed tank until it overflowed.

In this case readings were not ignored but simply not taken.

There were some errors in the stock sheets and the tank contained more than expected. However, if the operators had continued to dip the tank every shift the error would have been detected before the tank overflowed.

How can we prevent similar incidents happening again?

(1) Emphasize in operator training that they should take action on unusual readings, not just write them down. Have they been told what action to take?

(2) Mark control limits in red on record sheets. If readings are outside these limits, some action is required.

(3) Continue to take certain readings such as tank levels even when the plant is shut down. Tank levels are particularly liable to rise or fall when they should be steady.

3.3.2 WARNINGS IGNORED

When a warning is received, many operators are too ready to assume that the alarm is out of order. They thus ignore it or send for the instrument mechanic. By the time he confirms that the alarm is correct, it is too late. For example:

(a) During the morning shift, an operator noticed that a tank level was falling faster than usual. He reported that the level gauge was out of order and asked an instrument mechanic to check it. It was afternoon before he could do so. He reported that it was correct. The operator then looked around and found a leaking drain valve. Ten tons of material had been lost.

(b) Following some modifications to a pump, it was used to transfer some liquid. When the transfer was complete, the operator pressed the stop button on the control panel and saw that the "pump running" light went out. He also closed a remotely operated valve in the pump delivery line.

Several hours later the high-temperature alarm on the pump sounded. Because the operator had stopped the pump and seen the running light go out, he assumed the alarm was faulty and ignored it. Soon afterward there was an explosion in the pump.

When the pump was modified, an error was introduced into the circuit. As a result, pressing the stop button did not stop the pump but merely switched off the running light. The pump continued running, overheated, and the material in it decomposed explosively.

Operator training should emphasize the importance of responding to alarms. *They might be correct!*

If operators ignore alarms, it may be because experience has taught them that they are unreliable. Are your alarms adequately maintained?

3.3.3 IGNORANCE OF HAZARDS

In this section are a number of incidents which occurred because of ignorance of the most elementary properties of materials and equipment.

(a) A man who wanted some gasoline for cleaning decided to siphon it out of the tank of a company vehicle. He inserted a length of rubber tubing into the gasoline tank. Then, to fill the tubing and start the siphon, he held the hose against the suction nozzle of an industrial vacuum cleaner.

The gasoline caught fire. Two vehicles were destroyed and eleven damaged.

This occurred in a branch of a large organization, not a small company.

(b) A new cooler was being pressure-tested using a water pump driven by compressed air. A plug blew out, injuring the two men on the job.

It was then found that the pressure gauge had been fitted to the air supply instead of the cooler. Pressure had been taken far above the test pressure.

(c) An operator had to empty some tank trucks by gravity. He had been instructed to:
(1) Open the valve on top of the tank.
(2) Open the drain valve.
(3) When the tank was empty, close the valve on top of the tank.

He had to climb onto the top of the tank twice. He therefore decided to close the vent before emptying the tank. To his surprise, the tank was sucked in.

(d) At one plant it was discovered that contractors' employees were using welding cylinders to inflate pneumatic tires. The welders' torches made a good fit on the tire valves.

3.3.4 IGNORANCE OF SCIENTIFIC PRINCIPLES

The following incidents differ from those just described in that the operators, though generally competent, did not fully understand the scientific principles involved.

(a) A waste product had to be dissolved in methanol. The correct procedure was to put the waste in an empty vessel, box it up, evacuate it, break the vacuum with nitrogen and add methanol. When the waste had dissolved, the solution was moved to another vessel, the dissolving vessel evacuated again, and the vacuum broken with nitrogen.

If this procedure is followed, a fire or explosion is impossible because air and methanol are never in the vessel together.

However, to reduce the amount of work, the operators added the methanol as soon as the waste was in the vessel, without bothering to evacuate or add nitrogen. Inevitably, a fire occurred and a man was injured. As often happens, the source of ignition was never identified.

It is easy to say that the fire occurred because the operators did not follow the rules. But why did they not follow the rules? Perhaps because they did not understand that if air and a flammable vapor are mixed, an explosion may occur, and that we cannot rely on removing all sources of ignition. To quote from an official report, on a similar incident, "we do feel that operators' level of awareness about hazards to which they may be exposing themselves, has not increased at the same rate as has the level of personal responsibility which has been delegated to them" (3).

Also, the managers should have checked from time to time that the correct procedure was being followed.

(b) Welding had to take place near the roof of a storage tank which contained a volatile flammable liquid. There was a vent pipe on the roof of the tank, protected by a flame arrestor. Vapor coming out of this vent might have been ignited by the welding. The foreman therefore fitted a flex to the end of the vent pipe. The other end of the flex was placed on the ground so that the vapor now came out at ground level.

The liquid in the tank was soluble in water. As an additional precaution, the foreman therefore put the end of the flex in a drum of water. When the tank was emptied, the water first rose up the hose and then the tank was sucked in. The tank, like most such tanks, was designed for a vacuum of 2½ inches water gauge only (0.1 psi or 0.6 kPa) and would collapse at a vacuum of about 6 inches water gauge (0.2 psi or 1.5 kPa).

If the tank had been filled instead of emptied it might have burst because it was designed to withstand a pressure of only 8 inches water gauge (0.3 psi or 2 kPa) and would burst at about 3 times this pressure. Whether it burst or not would have depended on the depth of water above the end of the flex.

This incident occurred because the foreman, though a man of great experience, did not understand how a lute works. He did not realize how fragile storage tanks usually are. (See also Section 5.3.)

(c) The emergency blowdown valves in a plant were hydraulically operated and were kept shut by oil under pressure. One day the valves opened and the pressure in the plant blew off. It was then discovered that (unknown to the manager) the foremen, contrary to the instructions, were closing the oil supply valve "in case the pressure in the oil system failed"—a most unlikely occurrence and much less likely than the oil pressure leaking away from an isolated system.

Accidents which occurred because maintenance workers did not understand how things work or how they were constructed were described in Section 1.5.4.

3.3.5 ERRORS IN DIAGNOSIS

(a) The incident described in Section 3.2.8 is a good example of an error in diagnosis.

The operator correctly diagnosed that the rise in pressure in the reactor was due to a failure of the ethylene oxide to react. He decided that the temperature indicator might be reading high and that the temperature was therefore too low for reaction to start or that the reaction for some reason was sluggish to start and required a little more heat. He therefore raised the setting on the temperature trip and allowed the temperature to rise.

His diagnosis, though wrong, was not absurd. However, having made a diagnosis he developed a mind-set. That is, he stuck to it even though further evidence did not support it. The temperature rose but the pressure did not fall. Instead of looking for another explanation, or stopping the addition of ethylene oxide, he raised the temperature further and continued to do so until it reached 200°C instead of the usual 120°C. Only then did he realize that his diagnosis might be incorrect.

In developing a mind-set the operator was behaving like most of us. If we think we have found the solution to a problem, we become so committed to our theory that we close our eyes to evidence that does not support it. Specific training and practice in diagnostic skills (using, for example, the methods developed by Duncan and co-workers (4)) may make it less likely that operators will make errors in diagnosis.

(b) The accident at Three Mile Island in 1979 provided another example of an error in diagnosis (5). There were several indications that the level in the primary water circuit was low but two instruments indicated a high level. The operators believed these two readings and ignored the others. Perhaps their training had emphasized the hazard of too much water and the action to take but had not told them what to do if there was too little water in the system.

REFERENCES

1. J. Reason and K. Mycielska, *Absent Minded? The Psychology of Mental Lapses and Everyday Errors,* Prentice-Hall, 1982.
2. T. A. Kletz, *Chemical Engineering Progress,* Vol. 70, No. 4, April 1974, p. 80.
3. *Annual Report of Her Majesty's Inspectors of Explosives for 1970,* Her Majesty's Stationery Office, London, 1971.
4. E. E. Marshall and others, *The Chemical Engineer,* No. 365, Feb. 1981, p. 66.
5. T. A. Kletz, *Hydrocarbon Processing,* Vol. 61, No. 6, June 1982, p. 187.

Chapter 4

Labeling

Many incidents have occurred because equipment was not clearly labeled. Some have already been described in the section on the identification of equipment under maintenance (1.2).

Seeing that equipment is clearly and adequately labeled and checking from time to time to make sure that the labels are still there is a dull job, providing no opportunity to exercise our technical or intellectual skills. Nevertheless, it is as important as more demanding tasks. One of the signs of a good manager, foreman, operator, or designer is that he sees to the dull jobs as well as those that are fun. If you want to judge a team, look at their labels as well as the technical problems they have solved.

4.1 LABELING OF EQUIPMENT

(a) Small leaks of carbon monoxide from the glands of a compressor were collected by a fan and discharged outside the building. A man working near the compressor was affected by carbon monoxide. It was then found that a damper in the fan delivery line was shut. There was no label or other indication to show when the damper was closed and when it was open.

In a similar incident, a furnace damper was closed in error. It was operated pneumatically. There was no indication on the control knob to show which was the open position and which was the closed position.

(b) On several occasions it has been found that the labels on fuses or switchgear and the labels on the equipment they supply do not agree. The wrong fuses have then been withdrawn.

(c) Sample points are often unlabeled. As a result, the wrong material has often been sampled. This usually comes to light when the analysis results are received but sometimes a hazard develops. For example, a new employee took a sample of butane instead of a higher boiling liquid. The sample was placed in a refrigerator which became filled with vapor. Fortunately it did not ignite.

(d) Service lines are often not labeled. A fitter was asked to connect a steam supply at a gauge pressure of 200 psi (13 bar) to a process line to clear a choke. By mistake, he connected up a steam supply at a gauge pressure of 40 psi (3 bar). Neither supply was labeled and the 40 psi supply was not fitted with a nonreturn valve. The process material came back into the steam supply line.

Later, the steam supply was used to disperse a small leak. Suddenly the steam caught fire.

It is good practice to use a different type of connector on each type of service point.

(e) Two tank trucks were standing near each other in a filling bay. They were labeled as shown in Figure 4.1. The filler said to the drivers, "No.8 is ready." He meant that No.8 tank was ready, but the driver assumed that the tank attached to No.8 tractor was ready. He got into No.8 tractor and drove away. Tank No.4 was still filling.

Fortunately the tanker was fitted with a device to prevent it driving away when the filling hose was connected (1) and he was able to drive only a few yards.

If possible, tankers and tractors should be given entirely different sets of numbers.

(f) Nitrogen was supplied in tank cars which were also used for oxygen. Before filling the tank cars with oxygen, the filling connections were changed and hinged boards on both sides of the tanker were folded down so that they read "oxygen" instead of "nitrogen."

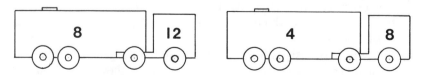

Figure 4.1. Arrangement of tank trailers and tractors.

A tank car was fitted with nitrogen connections and labeled "nitrogen." Probably due to vibration one of the hinged boards fell down, so that it read "oxygen." The filling station staff therefore changed the connections and put oxygen in it. Later, some nitrogen tank trucks were filled from the tank cars which were labeled "nitrogen" on the other side—and supplied to a customer who wanted nitrogen. He off-loaded the oxygen into his plant, thinking it was nitrogen (Figure 4.2).

The mistake was found when the customer looked at his weighbridge figures and noticed that on arrival the tanker had weighed three tons more than usual. A check then showed that the plant nitrogen system contained 30 percent oxygen.

Analyze all nitrogen tankers before off-loading.

4.2 LABELING OF INSTRUMENTS

(a) Plant pressures are usually transmitted from the plant to the control room by a pneumatic signal. This pneumatic signal, which is generated within the pressure sensing element, usually has a gauge pressure in the range 3 to 15 psi, covering the plant pressure from zero to maximum. For example, 3–15 psi (0.2 to 1 bar) might correspond to 0 to 1,200 psi plant pressure (0–80 bar).

The receiving gauge in the control room works on the transmitted pneumatic pressure, 15 psi giving full scale, but has its dial calibrated in terms of the plant pressure which it is indicating. The Bourdon tube of such a gauge is capable of withstanding only a limited amount of overpressure above 15 psi before it will burst. Furthermore, the material of the Bourdon tube is chosen for air and may be unsuitable for direct measurement of the process fluid pressure.

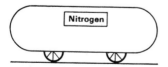 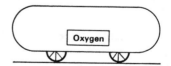

Figure 4.2. Arrangement of labels on tank cars. The "Nitrogen" label folds down to read "oxygen."

A pressure gauge of this sort, with a scale reading up to 1,200 psi, was installed directly in the plant. The plant gauge pressure was 800 psi and the gauge was damaged.

Gauges of this type should have the maximum safe working pressure clearly marked in red letters on the face.

(b) A workman, who was pressure testing some pipework with a hand-operated hydraulic pump, told his foreman that he could not get the gauge reading above 200 psi. The foreman told him to pump harder. He did, and burst the pipeline.

The gauge he was using was calibrated in atmospheres and not psi. The word "ats" was in small letters and in any case the workman did not know what it meant.

If more than one sort of unit is used in your plant for measuring pressure or any other property, then the units used should be marked on instruments in large, clear letters. You may use different colors for different units. Everyone should be aware of the differences between the units. However, it is better to avoid the use of different units.

(c) An extraordinary case of confusion between units occurred on a piece of equipment manufactured in Europe for a customer in England. The manufacturers were asked to measure all temperatures in °F and were told how to convert °C to °F.

A damper on the equipment was operated by a lever whose position was indicated by a scale, calibrated in degrees of arc. These were converted to °F!

(d) An operator was told to control the temperature of a reactor at 60°C. He set the set-point of the temperature controller at 60. The scale actually indicated 0–100 percent of a temperature range of 0–200°C so the set-point was really 120°C. This caused a runaway reaction which overpressured the vessel. Liquid was discharged and injured the operator (2).

(e) An error in testing made more probable by poor labeling was described in Section 3.2.4.

4.3 LABELING OF CHEMICALS

4.3.1 POOR OR MISSING LABELS

One incident was described in Section 2.8 (a).

Several incidents have occurred because drums or bottles were unlabeled and people assumed that they contained the material usually handled on the plant. In one case six drums of hypo (sodium hypochlorite) had to be added to a tank of water. Some of the drums were not labeled.

One contained sulphuric acid. It was added after some of the genuine hypo and chlorine was given off. The men adding the hypo were affected by the fumes.

In another case an unlabeled drum smelled like methylethylketone (MEK) so it was assumed to be MEK and was fed to the plant. Actually, it contained ethanol and a bit of MEK. Fortunately the only result was a ruined batch.

4.3.2 SIMILAR NAMES CONFUSED

Several incidents have occurred because similar names were confused. The famous case involving *Nutrimaster* (a food additive for animals) and *Firemaster* (a fire-retardant) is well known. The two materials were supplied in similar bags. A bag of *Firemaster,* delivered instead of *Nutrimaster,* was mixed into animal feeding stuffs causing an epidemic of illness among the farm animals (3).

Other chemicals that have been confused, with resultant accident or injury, are:

1. Washing soda (sodium carbonate) and caustic soda (sodium hydroxide)
2. Sodium nitrite and sodium nitrate
3. Sodium hydrosulphide and sodium sulphide
4. Ice and dry ice (solid carbon dioxide)
5. Photographers' hypo (sodium thiosulphate solution) and ordinary hypo (sodium hypochlorite solution).

In the last case a load of photographers' hypo was added to a tank containing the other sort of hypo. The two sorts of hypo reacted together, giving off fumes.

4.4 LABELS NOT UNDERSTOOD

Finally, the best labels are of no use if they are not understood.

(a) The word "slops" means different things to different people. A tank truck collected a load of slops from a refinery. The driver did not realize that they were flammable, took insufficient care and they caught fire. He thought "slops" were dirty water.
(b) A demolition contractor was required to use breathing apparatus while demolishing an old tank. He obtained several cylinders of compressed air, painted gray. Finding that they would be insufficient, he sent a truck for another cylinder. The driver returned with

a black cylinder. None of the men on the job, including the man in charge of the breathing apparatus, noticed the change or, if they did, attached any importance to it. When the new cylinder was brought into use a welder's face-piece caught fire. Fortunately he pulled it off at once and was not injured.

The black cylinder had contained oxygen.

All persons who are responsible for handling cylinders, particularly persons in charge of breathing apparatus, should be familiar with the color codes for cylinders.

REFERENCES

1. T. A. Kletz, *Loss Prevention,* Vol. 10, 1976, p. 151.
2. R. Fritz, *Safety Management* (S. Africa), Jan. 1982, p. 27.
3. *Business and Society,* Spring 1976, p. 5.

Chapter 5

Storage Tanks

No item of equipment is involved in more accidents than storage tanks, probably because they are fragile and easily damaged by slight overpressure or vacuum. Fortunately the majority of accidents involving tanks do not cause injury but they do cause damage, loss of material and interruption of production.

5.1 OVERFILLING

Most cases of overfilling are the result of lack of attention, wrong setting of valves, errors in level indicators and so on. See Section 3.3.1 (d). For this reason many companies fit high-level alarms to storage tanks. However, overfilling has still occurred because the alarms were not tested regularly or the warnings were ignored. See Section 3.3.2 (a).

Whether a high-level alarm is needed depends on the rate of filling and on the size of the batches being transferred into the receiving tank. If these are big enough to cause overfilling, a high level alarm is desirable.

Spillages resulting from overfilling should be retained in tank bunds. But very often the drain valves on the bunds—installed so that rain water can be removed—have been left open and the spillage is lost to drain. See Section 5.5.2 (c).

Drain valves should normally be locked shut. In addition, they should be inspected weekly to make sure that they are closed and locked.

5.1.1 ALARMS AND TRIPS CAN MAKE OVERFILLING MORE LIKELY

A high-level trip-on alarm may actually *increase* the frequency of overfilling incidents if its limitations are not understood.

At one plant a tank was filled each day with enough raw material for the following day. The operator watched the level. When the tank was full, he shut down the filling pump and closed the inlet valve. After several years, inevitably, one day he allowed his attention to wander and the tank overflowed. It was then fitted with a high-level trip which shut down the filling pump automatically.

To everyone's surprise the tank overflowed again a year later.

It had been assumed that the operator would continue to watch the level and that the trip would take over on the odd occasion when the operator failed to do so. Coincident failure of the trip was most unlikely. However, the operator no longer watched the level now that he was supplied with a trip. The manager knew that he was not doing so. But he decided that the trip was giving the operator more time for his other duties. The trip had the normal failure rate for such equipment, about once in two years, so another spillage after about two years was inevitable. A reliable operator had been replaced by a less reliable trip.

If a spillage about once in five years cannot be accepted, then it is necessary to have two protective devices, one trip (or alarm) to act as a process controller and another to take over when the controller fails. It is unrealistic to expect an operator to watch a level when a trip (or alarm) is provided. See Section 14.7 (a).

5.1.2 OVERFILLING DUE TO CHANGE OF DUTY

On more than one occasion tanks have overflowed because the contents were replaced by a liquid of lower specific gravity. The operators did not realize that the level indicator measured weight, not volume. For example, at one plant a tank which had contained gasoline (specific gravity 0.81) was used for storing pentane (specific gravity 0.69). The tank overflowed when the level indicator said it was only 85 percent full. The level indicator was a DP cell which measures weight.

Another incident is described in Section 8.2 (b).

If the level indicator measures weight it is good practice to fit a high level alarm which measures volume.

5.1.3 OVERFILLING BY GRAVITY

Liquid is sometimes transferred from one tank to another by gravity. Overfilling has occurred when liquid flowed from a tall tank to a shorter one. On one occasion an overflow occurred when liquid was transferred from one tank to another of the same height several hundred meters away. The operators did not realize that a slight slope in the ground was sufficient to cause the lower tank to overflow.

5.2 OVERPRESSURING

Most storage tanks are designed to withstand a gauge pressure of only eight inches of water (0.3 psi or 2 kPa) and will burst at about three times this pressure. They are thus easily damaged. Most (though not all) storage tanks are designed so that they will burst at the roof/wall weld, thus avoiding any spillage.

5.2.1 OVERPRESSURING WITH LIQUID

Suppose a tank is designed to be filled at a rate of x m³/hr. Many tanks, particularly those built some years ago, are provided with a vent big enough to pass x m³/hr of air but not x m³/hr of liquid. If the tank is overfilled, the delivery pump pressure will almost certainly be large enough to cause the tank to fail.

If the tank vent is not large enough to pass the liquid inlet rate then the tank should be fitted with a hinged manhole cover or similar overflow device. Proprietary devices are available.

This overflow device should be fitted to the roof near the wall. If it is fitted near the center of the roof, the height of liquid above the top of the walls may exceed eight inches and the tank may be overpressured. See Figure 5.1 (a).

Similarly, if the vent is designed to pass liquid it should be fitted near the edge of the roof and its top should not be more than eight inches above the top of the walls. Vessels have been overpressured because their vent pipes were too long. See Figure 5.1 (b). Tanks in which hydrogen

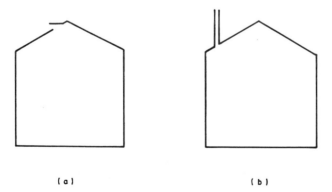

(a)   (b)

Figure 5.1. A tank may be overpressured if the vent or overflow is more than eight inches above the top of the walls.

may be evolved should be fitted with a vent at the highest point as well as an overflow. See Section 16.2.

5.2.2 OVERPRESSURING WITH GAS

This has usually occurred because those concerned did not realize that tanks are quite incapable of withstanding the pressure of the compressed air supply and that the vent may be too small to pass the inlet gas rate, as in the following two incidents:

(a) There was a choke on the exit line from a small tank. To try to clear the choke the operator held a compressed air hose against the open end at the top of the level glass. The gauge pressure of the compressed air was 100 psi (7 bar) and the top of the tank was blown off (Figure 5.2).
(b) An old vessel, intended for use as a low-pressure storage tank had been installed in a new position by a contractor who decided to pressure test it. He could not find a water hose to match the hose connection on the vessel and so he decided to use compressed air. The vessel ruptured.

   Another incident in which a storage vessel was ruptured by compressed air was described in Section 2.2 (a).
(c) On other occasions tanks have been ruptured because the failure of a level controller allowed a gas stream to enter the tank (Figure 5.3).

   The precautions necessary to prevent this occurring are analyzed in detail in Reference 1.

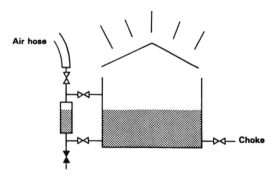

Figure 5.2. Tank top blown off by compressed air.

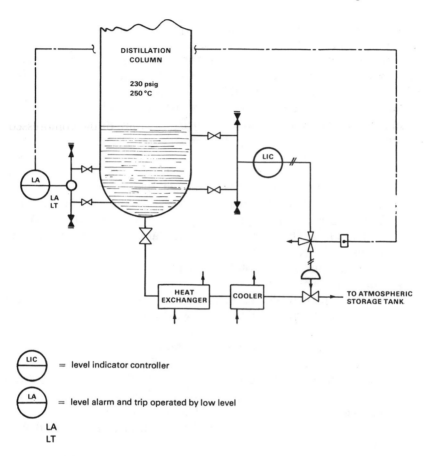

DISTILLATION
COLUMN

230 psig
250 °C

LIC

LA

LA
LT

HEAT
EXCHANGER

COOLER

TO ATMOSPHERIC
STORAGE TANK

LIC   = level indicator controller

LA   = level alarm and trip operated by low level

LA
LT

Figure 5.3. Showing how failure of a level controller can overpressure a tank.

5.3 SUCKING-IN

This is by far the commonest way in which tanks are damaged. The ways in which it occurs are legion. Some are listed below. Sometimes it seems that operators show great ingenuity in devising new ways of sucking in tanks!

Many of the incidents occurred because operators did not realize how fragile tanks are. They can be overpressured easily but sucked in much more easily. While most tanks are designed to withstand a gauge pressure of 8 inches of water (0.3 psi or 2 kPa) they are designed to withstand a vacuum of only $2^{1}/_{2}$ inches of water (0.1 psi or 0.6 kPa). This is the hydrostatic pressure at the bottom of a cup of tea.

Some incidents have occurred because operators did not understand how a vacuum works. See, for example, the incidents already described in Sections 3.3.3 (c) and 3.3.4 (b).

The following are some of the ways by which tanks have been sucked in. In some cases the vent was made ineffective. In others the vent was too small.

1. Three vents were fitted with flame arrestors which were not cleaned. After two years they choked. The flame arrestors were scheduled for regular cleaning (every six months) but this had been neglected due to pressure of work. See Section 6.2 (g).

   If you have flame arrestors on your tanks, are you sure they are necessary? See Section 6.2 (g).
2. A loose blank was put on top of the vent to prevent fumes coming out near a walkway.
3. After a tank had been cleaned a plastic bag was tied over the vent to keep dirt from getting in. It was a hot day. When a sudden shower cooled the tank, it collapsed.
4. A tank was boxed up with some water inside. Rust formation used up some of the oxygen in the air. See Section 11.1 (d).
5. While a tank was being steamed, a sudden thunderstorm cooled it so quickly that air could not be drawn in fast enough. When steaming out a tank, a manhole should be opened. Estimates of the vent area required range from 10 inches diameter to 20 inches diameter.

   On other occasions vent lines have been isolated too soon after steaming stopped. Tanks which have been steamed may require several hours to cool.
6. Cold liquid was added to a tank containing hot liquid.
7. A pressure/vacuum valve was assembled incorrectly—the pressure and vacuum pallets were interchanged. Valves should be designed so that this cannot occur. See Section 3.2.1.
8. A pressure/vacuum valve was corroded by the contents of the tank.
9. A larger pump was connected to the tank and it was pumped out more quickly than the air could get in through the vent.
10. Before emptying a tank truck the driver propped the manhole lid open. It fell shut.
11. A tank was fitted with an overflow which came down to ground level. There was no other vent. When the tank was overfilled the contents siphoned out (Figure 5.4).

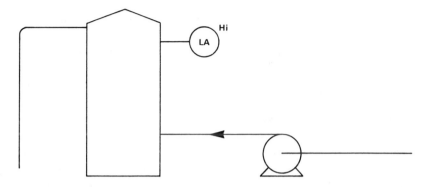

Figure 5.4. Overflow to ground level can cause a tank to collapse if there is no other vent.

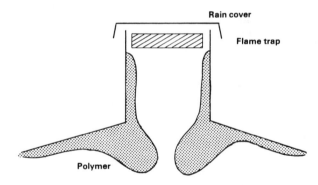

Figure 5.5. Vent almost blocked by polymer.

The tank should have been fitted with a vent on its roof, as well as the liquid overflow.

12. A vent was almost blocked by polymer (Figure 5.5). The liquid in the tank was inhibited to prevent polymerization but the vapor that condensed on the roof was not inhibited. The vent was inspected regularly but the polymer was not noticed.

Now a wooden rod is pushed through the vent to prove it is clear. (The other end of the rod should be enlarged so that it cannot fall into the tank.)

13. Water was added too quickly to a tank which had contained a solution of ammonia in water. To prevent the tank collapsing the vent would have had to be 30 inches in diameter! This is impractical, so the water should therefore be added slowly through a restriction orifice.

It is clear from these descriptions that we cannot prevent tanks being sucked in by writing lists of do's and don'ts or by altering plant designs, except in a few cases (see items 7 and 8). We can prevent these incidents only by increasing people's knowledge and understanding of the strength of storage tanks and of the way they work, particularly the way a vacuum works.

The need for such training is shown by the action taken following one of the incidents. Only the roof had been sucked in and it was concave instead of convex. The engineer in charge decided to blow the tank back to the correct shape by water pressure. He gave instructions for this to be done. A few hours later he went to see how the job was progressing. He found that the tank had been filled with water and that a hand-operated hydraulic pump, normally used for pressure testing pipework, was being connected to the tank. He had it removed and he had a vertical pipe, 1 m long, connected in place of the vent. He dribbled water into the pipe from a hose and as he did so the tank was restored to its original shape (Figure 5.6) to the amazement of onlookers.

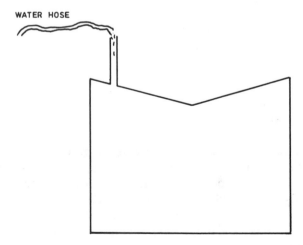

Figure 5.6. Method of restoring a tank with a concave roof to its original shape.

5.4 EXPLOSIONS

Explosions in the vapor spaces of fixed roof storage tanks have been numerous. One estimate puts the probability of an explosion at about once in 1,000 years per tank, based on historical records. The reason for the large number of explosions is that explosive mixtures are present in the vapor spaces of many storage tanks. It is almost impossible to be certain that a source of ignition will never turn up, particularly if the liquid in the tank has a low conductivity so that static charges can accumulate on the liquid. For this reason many companies insist that fixed roof storage tanks containing hydrocarbons above their flash points must be blanketed with nitrogen. Other companies insist that such hydrocarbons are stored only in floating roof tanks.

Non-hydrocarbons, even if flammable, have a higher conductivity than hydrocarbons. Charges of static electricity can rapidly drain away to earth (provided the equipment is earthed) and the risk of ignition is much lower. Many companies therefore store these materials in fixed roof tanks without nitrogen blanketing (2).

5.4.1 A TYPICAL TANK EXPLOSION

A large tank blew up 40 minutes after the start of a blending operation in which one grade of naphtha was being added to another. The fire was soon put out and the naphtha was moved to another tank.

The next day blending was resumed; 40 minutes later another explosion occurred.

The tanks were not nitrogen blanketed and there was an explosive mixture of naphtha vapor and air above the liquid in the tanks. The source of ignition was static electricity. The pumping rate was rather high so that the naphtha flowing through the pump and lines acquired a charge. A spark passed between the liquid in the tank and the roof or walls of the tank, igniting the vapor-air mixture.

These explosions led to an extensive series of investigations into the formation of static electricity (3).

There are several ways of preventing similar explosions.

1. Use nitrogen blanketing or floating roof tanks.
2. Use anti-static additives; they increase the conductivity of the liquid so that charges can drain away rapidly to earth (provided equipment is earthed). However, make sure that the additives do not deposit on catalysts or interfere with chemical operations in other ways.
3. Minimize the formation of static electricity by keeping pumping rates low (less than 3 m/s for pure liquids but less than 1 m/s if

water is present) and avoiding splash filling. Filters and other restrictions should be followed by a long length of straight line to allow charges to decay.

It is difficult to feel confident that (3) can always be achieved and therefore (1) or (2) are recommended.

For more information on static electricity see Chapter 15.

5.4.2 SOME UNUSUAL TANK EXPLOSIONS

(a) A new tank was being filled with water for hydrostatic test when an explosion occurred. Two welders who were working on the roof, finishing the hand-rails, were injured, fortunately not seriously.

The tank had been filled with water through a pipeline that had previously contained gasoline. A few liters left in the line were flushed into the tank by the water and floated on top of it. The vapor was ignited by the welders.

No one should be allowed to go onto the roof of a tank while it is being filled with water for testing. One of the reasons for filling it with water is to make sure that the tank and its foundations are strong enough. People should be kept out of the way, in case they are not.

(b) During the construction of a new tank the contractors decided to connect the nitrogen line to the tank. They knew better, they said, than to connect the process lines without authority. But nitrogen was inert and therefore safe.

The new tank and an existing one were designed to be on balance with each other to save nitrogen (Figure 5.7), but the contractors did not understand this.

The valve to the new tank was closed but leaking. Nitrogen and methanol vapor entered the tank and the vapor was ignited by a welder who was completing the inlet line to the tank. The roof was blown right off. By great good fortune, it landed on a patch of empty ground just big enough to contain it (Figure 5.8).

(c) The roof of an old gasoline tank had to be repaired. The tank was steamed out and cleaned and tests with a combustible gas detector showed that no flammable gas or vapor was present.

A welder was therefore allowed to start work. Soon afterward a small flash of flame singed his hair.

The roof was made from plates which overlapped each other by about four inches and which were welded together on the top side

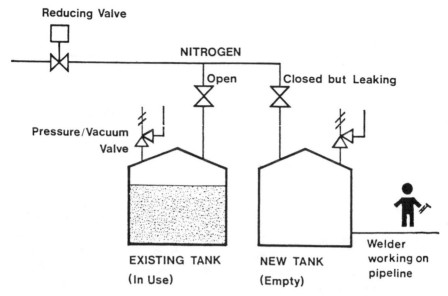

Figure 5.7. If tanks are on balance the nitrogen entering one tank is inevitably mixed with vapor.

Figure 5.8. When an explosion occurred in a tank, the roof landed on an area just big enough to contain it.

Figure 5.9. An old method of tank construction allows liquid to enter the gap between the plates.

only—an old method of construction not now used (Figure 5.9).

It is believed that some gasoline entered the space between the plates and became trapped by rust and scale. The heat from the welding vaporized the gasoline and it blew out of the molten weld.

At the time, the suggestion was made that the tank should be filled with water but this cannot be done without risking over-pressuring the tank. See Section 5.2.1.

Fires and explosions which occurred while repairing or demolishing storage tanks containing traces of heavy oil are described in Section 12.4.1.

5.4.3 AN EXPLOSION IN AN OLD PRESSURE VESSEL USED AS A STORAGE TANK

Sometimes old pressure vessels are used as storage tanks. It would seem that by using a stronger vessel than is necessary, we achieve greater safety. But this may not be the case, as if the vessel fails, it will do so more spectacularly. See Section 2.2 (a).

A tank truck hit a pipeline leading to a group of tanks. The pipeline went over the top of the bund wall and it broke off inside the bund. The engine of the truck ignited the spillage, starting a bund fire which damaged or destroyed 21 tanks and 5 tank trucks.

An old 100 m^3 pressure vessel, a vertical cylinder, designed for a gauge pressure of 5 psi (0.3 bar) was being used to store, at atmospheric pressure, a liquid of flash point 40°C. The fire heated the vessel to above 40°C and ignited the vapor coming out of the vent; the fire flashed back into the tank where an explosion occurred. The vessel burst at the bottom seam and the entire vessel, except for the base, and contents went into orbit like a rocket (4).

If the liquid had been stored in an ordinary low-pressure storage tank, then the roof would have come off and the burning liquid would have been retained in the rest of the tank.

The incident also shows the importance of cooling, with water, all tanks or vessels exposed to fire. It is particularly important to cool vessels. They fail more catastrophically, either by internal explosion or because the rise in temperature weakens the metal. See Section 8.1.

Another tank explosion is described in Section 16.2 (a).

5.5 FLOATING ROOF TANKS

This Section describes some incidents which could only have occurred on floating roof tanks.

5.5.1 HOW TO SINK THE ROOF

The flexible roof drain on a floating roof tank choked, so it was decided to drain rainwater off the roof with a hose.

To prime the hose and establish a siphon, the hose was connected to the water supply. It was intended to open the valve on the water supply for just long enough to fill the hose. This valve would then be closed and the drain valve opened (Figure 5.10). However, the water valve was opened

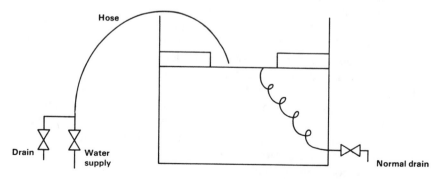

Figure 5.10. How to sink the roof of a floating roof tank.

in error and left open, with the drain valve shut. Water flowed onto the floating roof and it sank in thirty minutes.

Temporary modifications should be examined with the same thoroughness as permanent ones. See Section 2.4.

5.5.2 FIRES AND EXPLOSIONS

(a) Most fires on floating roof tanks are small rim fires caused by vapor leaking through the seals. The source of ignition is often atmospheric electricity. It can be eliminated as a source of ignition by fitting shunts—strips of metal—about every meter or so around the rim to ground the roof to the tank walls.

Many rim fires have been extinguished by a worker using a hand held fire extinguisher. However, in 1979 a rim fire had just been extinguished when a pontoon compartment exploded, killing a fireman. It is believed that there was a hole in the pontoon and some of the liquid in the tank leaked into it.

Workers should not go onto floating roof tanks to extinguish rim fires (5). If fixed fire-fighting equipment is not provided, foam should be supplied from a monitor.

(b) The roof of a floating roof tank had to be replaced. The tank was emptied, purged with nitrogen and steamed for six days. Each of the float chambers was steamed for four hours. Rust and sludge were removed from the tank. Demolition of the roof was then started.

Fourteen days later a small fire occurred. About a gallon of gasoline came out of one of the hollow legs which support the roof when it is off float, and was ignited by a spark. The fire was put out with dry powder.

It is believed that the bottom of the hollow leg was blocked with sludge and that as cutting took place near the leg it moved and disturbed the sludge (Figure 5.11).

Before welding or burning is permitted on floating roof tanks, the legs should be flushed with water from the top.

On some tanks the bottoms of the legs are sealed. Holes should be drilled in them so that they can be flushed through.

(c) Sometimes a floating roof is inside a fixed roof tank. In many cases this will reduce the concentration of vapor in the vapor space below the explosive limit. But in other cases it can increase the hazard because vapor which was previously too rich to explode is brought into the explosive range.

A serious fire which started in a tank filled with an internal floating roof is described in Reference 6.

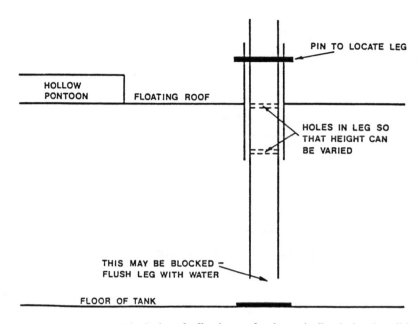

Figure 5.11. Oil trapped in the leg of a floating roof tank caught fire during demolition.

As a result of a late change in design the level at which a floating roof came off-float had been raised but this was not marked on the drawings which were given to the operators. As a result, without intending to, they took the roof off-float. The pressure-vacuum valve opened, allowing air to be sucked into the space beneath the floating roof.

When the tank was refilled with warm crude oil at 37°C, vapor was pushed out into the space above the floating roof and then out into the atmosphere through vents on the fixed roof tank (Figure 5.12).

This vapor was ignited at a boiler house some distance away. The fire flashed back to the storage tank and the vapor burned as it came out of the vents.

Pumping was therefore stopped. Vapor no longer came out of the vents, air got in and a mild explosion occurred inside the fixed roof tank. This forced the floating roof down like a piston and some of the crude oil came up through the seal past the side of the floating roof and out of the vents on the fixed roof tank. This oil caught

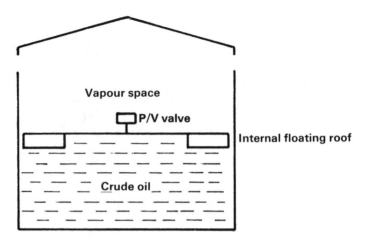

Figure 5.12. Tank with internal floating roof.

fire, causing a number of pipeline joints to fail and this caused further oil leakages. One small tank burst; fortunately it had a weak seam roof.

Over fifty fire appliances and 200 firemen attended and the fire was under control in a few hours.

The water level outside the bund rose because the bund drain valve had been left open and the bund wall was damaged by the fire-fighting operations. The firemen pumped some of the water into another bund, but it ran out because the drain valve on this bund had also been left open.

An overhead power cable was damaged by the fire and fell down, giving someone an electric shock. The refinery staff therefore isolated the power to all the cables in the area. Unfortunately they did not tell the firemen what they were going to do. Some electrically driven pumps which were pumping away some of the excess water stopped and the water level rose even further. Despite a foam cover, oil floating on top of the water was ignited by a fire engine which was standing in the water. The fire spread rapidly for 150 m. Eight firemen were killed and two seriously injured. A naphtha tank ruptured, causing a further spread of the fire and it took 15 hours to bring it under control.

The main lessons from this incident are:

1. Keep plant modifications under control and keep drawings up to date. See Chapter 2.
2. Do not take floating roof tanks off-float except when they are being emptied for repair.
3. Keep bund drain valves locked shut. Check regularly to make sure they are shut.
4. Plan now how to get rid of fire-fighting water. If the drains will not take it, it will have to be pumped away.
5. During a fire keep in close touch with the firemen and tell them what you propose to do.

5.6 MISCELLANEOUS INCIDENTS

5.6.1 A TANK RISES OUT OF THE GROUND

A tank was installed in a concrete-lined pit. The pit was then filled with sand and a layer of concrete six inches thick was put over the top. Water accumulated in the pit and the buoyancy of the tank was sufficient to break the holding-down bolts and push it through the concrete covering.

A sump and pump had been provided for the removal of water. But either the pump-out line had become blocked or pumping had not been carried out regularly (7).

Underground tanks are not recommended for plant areas. They cannot be inspected for external corrosion and the ground is often contaminated with corrosive chemicals.

5.6.2 A CORROSION FAILURE

Part of the sand foundation beneath a 12-year-old tank subsided. Water collected in the space that was left and caused corrosion. This was not detected because the insulation on the tank came right down to the ground.

When the corrosion had reduced the wall thickness from 6 mm to 2 mm, the floor of the tank collapsed along a length of 2.5 m, and 30,000 m^3 of hot fuel oil came out. Most of it was collected in the bund. However, some leaked into other bunds through rabbit holes in the earth walls.

All storage tanks should be scheduled for inspection every few years. And on insulated tanks the insulation should finish 200 mm above the base so that checks can be made for corrosion.

5.6.3 NITROGEN BLANKETING

Section 5.4.1 discussed the need for nitrogen blanketing. However, if it is to be effective, it must be designed and operated correctly.

(a) INCORRECT DESIGN

On one group of tanks the reducing valve on the nitrogen supply was installed at ground level (Figure 5.13). Hydrocarbon vapor condensed in the vertical section of the line and effectively isolated the tank from the nitrogen blanketing.

The reducing valve should have been installed at roof height.

Check your tanks—there may be some more like this one.

(b) INCORRECT OPERATION

An explosion and fire occurred on a fixed roof tank which was supposed to be blanketed with nitrogen. After the explosion it was found that the nitrogen supply had been isolated. Six months before the explosion the manager had personally checked that the nitrogen blanketing was in operation. But no later check had been carried out (8).

All safety equipment and systems should be scheduled for regular inspection and test. Nitrogen blanketing systems should be inspected at least weekly. It is not sufficient to check that the nitrogen is open to the tank. The atmosphere in the tank should be tested with a portable oxygen analyzer to make sure that the oxygen concentration is below 5 percent.

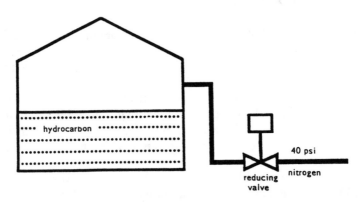

Figure 5.13. Incorrect installation of nitrogen blanketing.

Large tanks (say, over 1,000 m^3) blanketed with nitrogen should be fitted with low pressure alarms to give immediate warning of the loss of nitrogen blanketing.

REFERENCES

1. T. A. Kletz and H. G. Lawley, in A. E. Green (editor), *High Risk Safety Technology*, Wiley, 1982, Chapter 2.1.
2. T. A. Kletz, *Institution of Chemical Engineers Symposium Series No. 34*, 1971, p. 75.
3. A. Klinkenberg and J. L. van der Minne, *Electrostatics in the Petroleum Industry*, Elsevier, Amsterdam, 1958.
4. *Loss Prevention*, Vol. 7, 1972, p. 119 and Case History No. 1887, Manufacturing Chemists Association, Washington, D.C.
5. D. K. McKibben, *Safe Design of Atmospheric Pressure Vessels*, Seminar on Prevention of Fires and Explosions in the Hydrocarbon Industries, 21–26 June 1982, Institute of Gas Technology, Chicago.
6. Press release issued by the City of Philadelphia Office of the City Representatives, 12 Dec. 1975.
7. *Petroleum Review*, Oct. 1974, p. 683.
8. T. A. Kletz, *Chemical Engineering Progress*, Vol. 70, No. 4, April 1974, p. 80.

Chapter 6

Stacks

Stacks, like storage tanks are, or have been, the sites of numerous explosions. They have also been known to choke.

6.1 STACK EXPLOSIONS

(a) Figure 6.1 shows the results of an explosion in a large flare stack. The stack was supposed to be purged with inert gas. However, the flow was not measured and had been cut back almost to zero to save nitrogen. Air leaked in through the large bolted joint between unmachined surfaces. The flare had not been alight for some time. Shortly after it was relit, the explosion occurred—the next time some gas was put to stack. The mixture of gas and air moved up the stack until it was ignited by the pilot flame.

To prevent similar incidents happening again:

1. Stacks should be welded. They should not contain bolted joints between unmachined surfaces.
2. There should be a continuous flow of gas up every stack to prevent air diffusing down and to sweep away small leaks of air into the stack. The continuous flow of gas does not have to be nitrogen—a waste-gas stream is effective. But if gas is not being flared continuously it is usual to keep nitrogen flowing at a linear velocity of 0.03–0.06 m/sec. The flow of gas should be measured. A higher rate is required if hydrogen or hot condensible gases are being flared. If possible, hydrogen should be discharged through a separate vent stack and not mixed with other gases in a flare stack.

Figure 6.1. Base of flare stack.

3. The atmosphere inside every stack should be monitored regularly, say daily, for oxygen content. Large stacks should be fitted with oxygen analyzers which alarm at 5 percent (2 percent if hydrogen is present). Small stacks should be checked with a portable analyzer.

   These recommendations apply to vent stacks as well as flare stacks.

(b) Despite the publicity given to the incident just described, another stack explosion occurred nine months later in the same plant.

To prevent leaks of carbon monoxide and hydrogen from the glands of a number of compressors getting into the atmosphere of the compressor house, they were sucked away by a fan and discharged through a small vent stack. Air leaked into the duct because there was a poor seal between the duct and the compressor. The mixture of air and gas was ignited by lightning.

The explosion would not have occurred if the recommendations made after the first explosion had been followed—if there had been a flow of inert gas into the vent collection system and if the atmosphere inside had been tested regularly for oxygen.

Why were they not followed? Perhaps because it was not obvious that recommendations made following an explosion on a large flare stack applied to a small vent stack.

(c) Vent stacks have been ignited by lightning or in other ways on many occasions. On several occasions a group of ten or more stacks have been ignited simultaneously. This is not dangerous provided that:

1. The gas mixture in the stack is not flammable so that the flame cannot travel down the stack.
2. The flame does not impinge on overhead equipment. (Remember that in a wind it may bend at an angle of 45°.)
3. The flame can be extinguished by isolating the supply of gas or by injecting steam or an increased quantity of nitrogen. (The gas passing up the stack will have to contain over 90 percent nitrogen to prevent it forming a flammable mixture with air.)

Other stack explosions have been described by Kilby (1).

6.2 BLOCKED STACKS

(a) Section 2.5 (a) described how an 8-inch-diameter vent stack became blocked by ice because cold vapor ($-100°C$) and steam were passed up the stack together. The cold gas met the condensate running down the walls and caused it to freeze. A liquefied gas tank was overpressured and a small split resulted. The stack was designed to operate without steam. But the steam was then introduced to make sure that the cold gas dispersed and did not drift down to ground level.

(b) The vent stack was replaced by a 14-inch-diameter flare stack with a supply of steam to a ring round the top of the stack. A few years later this stack also choked again, this time due to a deposit of refractory debris from the tip, cemented together by ice (as some condensate from the steam had found its way down the stack). Fortunately, in this case the high pressure in the tank was noticed before any damage occurred. There was no boot at the bottom of the stack to collect debris (Figure 6.2). A boot was fitted (2).

(c) On other occasions blowdown lines or stacks have become blocked in cold weather because benzene or cyclohexane, both of which have freezing points of 5°C, were discharged through them. Steam tracing of the lines or stacks may be necessary.

(d) Blowdown lines should never be designed with a dip in them or liquid may accumulate in the dip and exert a back pressure. This has caused vessels to be overpressured (3).

(e) A blowdown line which was not adequately supported sagged when exposed to fire and caused a vessel to be overpressured.

(f) Water seals have frozen in cold weather. They should not be used except in locales where freezing cannot occur.

Flare and vent systems should be simple. It is better to avoid water seals than install steam heating systems and low temperature alarms which might fail.

(g) Vent stacks are sometimes fitted with flame arrestors, to prevent a flame on the end of the stack traveling back down the stack. The arrestors are liable to choke unless regularly cleaned. They are also unnecessary because unless the gas mixture in the stack is flammable, the flame cannot travel down the stack. If the gas mixture in the stack is flammable, then it may be ignited in some other

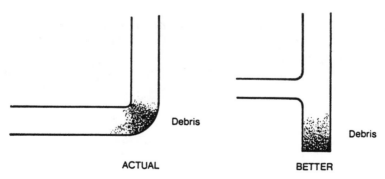

ACTUAL   BETTER

Figure 6.2. Flare stack after explosion.

way. Stacks should therefore be swept by a continuous flow of gas to prevent a flammable mixture forming, as discussed in Section 6.1.

There are, however, two cases in which flame arrestors in vent stacks are justified:

1. If the gas being vented can decompose without the addition of air. Whenever possible such gases should be diluted with nitrogen. If this is not always possible a flame trap may be used.
2. In the vent pipes of storage tanks containing a flammable mixture of vapor and air (Section 5.4.1). Such flame traps should be inspected regularly and cleaned if necessary. Section 5.3 (1) described how a tank was sucked in because the flame arrestors on all three vents had not been cleaned for two years.

   One type of flame arrestor that can be easily removed for inspection without the need to use tools is described in Reference 4.

(h) Molecular seals have been choked by carbon from incompletely burned gas. For this reason, many companies prefer not to use them.

The common theme of many of these items is that blowdown lines, flare and vent stacks should be kept simple because they are part of the pressure relief system. Avoid flame arrestors, molecular seals, water seals and U-bends. Avoid steam, which brings with it rust and scale and may freeze.

REFERENCES

1. J. L. Kilby, *Chemical Engineering Progress,* June 1968, p. 419.
2. T. A. Kletz, *Chemical Engineering Progress,* Vol. 70, No. 4, April 1974, p. 80.
3. T. J. Laney in *Fire Protection Manual for Hydrocarbon Processing Plants,* edited by C.H. Vervalin, Vol. 1, 3rd Edition, Gulf Publishing Company, 1985, p. 101.
4. T. A. Kletz, *Plant/Operations Progress,* Vol. 1, No. 4, October 1982, p. 252.

Chapter 7

Leaks

Leaks of process materials are the process industries' biggest hazard. Most of the materials handled will not burn or explode unless mixed with air in certain proportions. To prevent fires and explosions we must therefore keep the fuel in the plant and the air out of the plant. The latter is relatively easy because most plants operate at pressure. Nitrogen is widely used to keep air out of low-pressure equipment such as storage tanks (Section 5.4), stacks (Section 6.1), centrifuges (Section 10.1) and equipment which is depressured for maintenance (Section 1.3).

The main problem in preventing fires and explosions is thus preventing the process material leaking out of the plant, that is, maintaining plant integrity. Similarly, if toxic or corrosive materials are handled, they are hazardous only when they leak.

Many leaks have been discussed under other headings—leaks which occurred during maintenance (Chapter 1), as the result of human error (Chapter 3) or as the result of overfilling storage tanks (Section 5.1). Other leaks have occurred as the result of pipe or vessel failures (Chapter 9), while leaks of liquefied flammable gas are discussed in Chapter 8.

Here, we discuss some other sources of leaks and the isolation and control of the leaking material.

7.1 SOME COMMON SOURCES OF LEAKS

7.1.1 SMALL COCKS

Small cocks have often been knocked open or have vibrated open. They should never be used as the sole isolation valve (and preferably not

93

at all) on lines carrying hazardous materials, particularly flammable or toxic liquids, at pressures above their atmospheric boiling points (for example, liquefied flammable gases or most heat transfer oils when hot). These liquids turn to vapor and spray when they leak and can spread long distances.

It is good practice to use other types of valves for the first isolation valve, as shown in Figure 7.1.

7.1.2 DRAIN VALVES AND VENTS

Many leaks have occurred because workers left drain valves open while draining water from storage tanks or process equipment and then returned to find that oil was running out instead of water.

In one incident a man was draining water from a small distillation column rundown tank containing benzene. He left the water running for a few minutes to attend to other jobs. Either there was less water than usual or he was away longer than expected. He returned to find benzene running out of the drain valve which was about two inches in diameter. Before he could close it, the benzene was ignited by the furnace which heated the distillation column. The operator was badly burned and died from his injuries.

The furnace was too near the drain point (it was about 10 m away) and the slope of the ground allowed the benzene to spread towards the furnace. Nevertheless, the fire would not have occurred if the drain valve had not been left unattended.

Spring-loaded ball valves should be used for drain valves. They have to be held open, and close automatically if released.

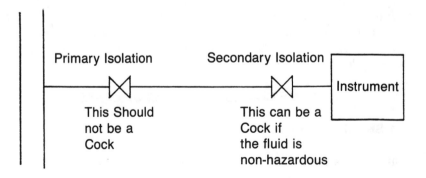

Figure 7.1. Small cocks should not be used as primary isolation valves.

The size of drain valves should be kept as small as practicable. With liquefied flammable gases and other flashing liquids, ¾-inch should be the maximum allowed.

Drain valves which are used only occasionally to empty equipment for maintenance should be blanked when not in use. Regular surveys should be made to see that the blanks are in position. On one plant a survey after a turnaround showed that 50 blanks were loose, each hanging on one bolt.

If water has to be drained regularly from liquefied flammable gases or other flashing liquids, and if a spring-loaded valve cannot be used, then a remotely operated emergency isolation valve (see Section 7.2.1) should be installed in the drain line.

While drain valves are installed to get rid of unwanted liquid, vent lines get rid of unwanted gas or vapor. They should be located so that the vapor is unlikely to ignite, so that damage is minimal if it does ignite and so that people are not affected by the gas or vapor discharged. One fire destroyed a small plant. It started because the vent on a distillation column condenser discharged into the control room (1).

7.1.3 OPEN CONTAINERS

Buckets and other open-topped containers should never be used for collecting drips of flammable, toxic or corrosive liquids or for carrying small quantities about the plant. Drips, reject samples, etc., should be collected in closed metal cans and the caps should be fitted before the cans are moved.

One man was badly burned when he was carrying gasoline in a bucket and it caught fire. The source of ignition was never found. Another man was carrying phenol in a bucket when he slipped and fell. The phenol spilled onto his legs. Half-an-hour later he was dead. A third man was moving a small open-topped drum containing hot cleaning fluid. He slipped; liquid splashed onto him and scalded him.

Other incidents are described in Sections 12.2 (c) and 15.1.

These incidents may seem trivial compared with those described in other pages. But for the men concerned they were their Flixborough.

Similarly, glass sample bottles should never be carried by hand. Workers have been injured when bottles they were carrying knocked against projections and broke. Bottles should be carried in baskets or other containers such as those used for soft drinks. Bottles containing particularly hazardous chemicals such as phenol should be carried in closed containers.

Flammable liquids should, of course, never be used for cleaning floors or for cleaning up spillages of dirty oil. Nonflammable solvents should be used.

7.1.4 LEVEL GLASSES

Failures of level glasses and sight glasses have caused many serious incidents. A leak of ethylene and an explosion which destroyed a plant may have been due to a level glass failure (2).

Level glasses and sight glasses should not be used on vessels containing flashing flammable or toxic liquids—that is, liquids under pressure above their normal boiling points. When level glasses are used they should be fitted with ball check cocks which prevent a massive leak if the glass breaks. Unfortunately, the balls have sometimes been removed by people who did not understand their purpose. The hand valves must be fully open or the balls cannot operate (Figure 7.2).

7.1.5 PLUGS

On many occasions screwed plugs have blown out of equipment.

(a) A ½-inch plug was fitted in a bellows so that after pressure testing, in a horizontal position, water could be completely drained out. Soon after the bellows was installed the plug blew out followed by a jet of hot oil 30 m long.

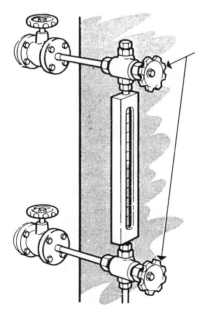

These cocks contain a ball which will isolate the sight glass if the glass breaks.

The cocks must be FULLY OPEN to allow the ball to seat if the glass breaks.

NEVER REMOVE THE BALLS. CHECK THAT THEY ARE THERE.

Figure 7.2. Ball check cocks.

Plugs installed to facilitate pressure testing should be welded in position. However, it is bad practice to seal weld over an ordinary screwed plug. If the thread corrodes, the full pressure is applied to the seal. A specially designed plug with a full-strength weld should be used.

(b) A 1-inch plug blew out of a pump body followed by a stream of oil at 370°C and a gauge pressure of 250 psi (17 bar). The oil caught fire and caused extensive damage. The plug had been held by only one or two threads and had been in use for 18 years.

Following this incident, surveys at other plants brought to light many other screwed plugs, some held by only a few threads and some made from the wrong grade of steel. At one plant, which did not allow the use of screwed plugs, several 2-inch plugs were found, held by only one thread. They had been in use for 10 years and were supplied as part of a compressor package.

A survey of all plugs is recommended.

(c) A similar incident is described in Section 9.1.6 (e). A screwed nipple and valve, installed for pressure testing, blew out of an oil line.

(d) The hinge-pin retaining plug on a standard swing nonreturn valve worked loose and blew out. Gas leaked out at a rate of 2 t/h until the plant could be shut down.

This incident emphasizes the point made in Section 7.2.1 (b). Nonreturn valves have a bad name among many plant operators but no item of equipment can be expected to function correctly if it is never maintained.

(e) A valve was being overhauled in a workshop. A screwed plug was stuck in the outlet. To loosen the plug the valve was heated with a welding torch. It shattered.

The valve was in the closed position and some water was trapped between the valve and the plug.

Valves should normally be opened before they are maintained.

7.1.6 HOSES

Hoses are a frequent source of leaks. The most common reasons have been:

1. A hose made of the wrong material was used.
2. The hose was damaged.
3. The connections were not made correctly. In particular, screwed joints were secured by only a few threads, or different threads were combined or gaskets were missing.

4. The hose was fixed to the connector or to the plant by a screwed clip of the type used for automobile hoses (Jubilee clips). These are unsuitable for industrial use. Bolted clamps should be used.
5. The hose was disconnected before the pressure had been blown off, sometimes because there was no vent valve through which it could be blown off.
6. The hose was used for a service such as steam or nitrogen. The service valve was closed before the process valve. As a result, process materials entered the hose.

These points are illustrated by the following incidents:

(a) It was decided to inject live steam at a gauge pressure of 100 psi (7 bar) into a distillation column to see if this improved its performance. An operator was standing in the position shown in Figure 7.3 and was about to close the inlet valve to the column when the hose burst. He was showered with hot, corrosive liquid. He was standing on an access platform. The leak prevented him from reaching the access ladder. He had to wait until someone fetched a portable ladder.
The investigation showed that:

1. The hose was made of reinforced rubber, the wrong material. A stainless steel hose should have been used.
2. The hose was damaged.
3. The steam valve at the other end of the hose was closed just before the column inlet valve, thus allowing process material to enter the hose. The operators knew this was not normal practice. But they closed the steam valve first because they knew the hose was damaged and wanted to avoid subjecting it to the full steam pressure.

The right type of hose should have been used, it should have been in good condition, and the process valve should have been closed first. In addition, a valve on a hose should not be in a position to which access is so poor. If no other valve was available, a steel pipe should have been fitted to the valve so that the end of the hose was in a safer place.

All hoses should be inspected and tested regularly and marked to show that they have been approved for use. A good practice is to change the color of the label every 6 or 12 months. This incident is a good illustration of the way in which both operators and managers become so used to the hazards of process materials that they fail

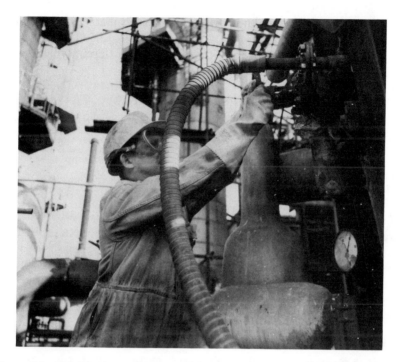

Figure 7.3. This hose burst, injuring the operator. It was the wrong type, was damaged and was badly located.

to establish and maintain proper precautions. How often had the wrong hose or a damaged hose been used before? Why had the foremen or the managers not noticed them?

(b) A tank truck containing 60 percent oleum arrived at a plant. The truck's hose was damaged, so the operators found a hose which was intended for use with 20 percent oleum. After 45 minutes it leaked and there was a large spillage. The operators assumed that the hose must have been damaged. They replaced it with a similar one and after 15 minutes another spillage occurred.

This incident illustrates the "mind-sets" described in Section 3.3.5. Having assumed that a hose used for 20 percent oleum would be suitable for any sort of oleum, the operators stuck to their opinion even though the hose leaked. They thus had an "action replay."

Do your operators know which hoses are suitable for which materials?

Mistakes are less likely if the number of different types used is kept to a minimum.

Other hose failures are described in Section 13.2.

7.2 CONTROL OF LEAKS

7.2.1 EMERGENCY ISOLATION VALVES (EIVs)

Many fires have been prevented or quickly extinguished by remotely operated emergency isolation valves. We cannot install them in the lines leading to all equipment which *might* leak. However, we can install them in the lines leading to equipment which experience shows is particularly liable to leak (for example, very hot or cold pumps or drain lines, as described in Section 7.1.2) or in lines from which, if a leak did occur, a very large quantity of material, say 50 tons or more, would be spilled (for example, the bottoms pumps or reflux pumps on large distillation columns).

In all these cases, once the leak starts, particularly if it ignites, it is usually impossible to approach the normal hand-isolation valves to close them. Emergency isolation valves are discussed in detail in Reference (3) and the following incidents show how useful they can be. They can be operated electrically, pneumatically, or in some cases, hydraulically.

(a) A leak of light oil from a pump caught fire. The flames were 10 m high. From the control room, the operator closed a remotely operated valve in the pump suction line. The flames soon died down and the fire burned itself out in 20 minutes. It would have been impossible to have closed a hand-operated valve in the same position. And if the emergency valve had not been provided, the fire would have burned for many hours. The emergency valve had been tested regularly. It could not be fully closed during testing but was closed part-way.

Backflow from the delivery side of the pump was prevented by a nonreturn valve. In addition, a control valve and a hand valve well away from the fire were closed (Figure 7.4).

(b) The bearing on the feed pump to a furnace failed, causing a gland failure and a leak of hot oil. The oil caught fire, but an emergency isolation valve in the pump suction line was immediately closed and the fire soon died out (Figure 7.5).

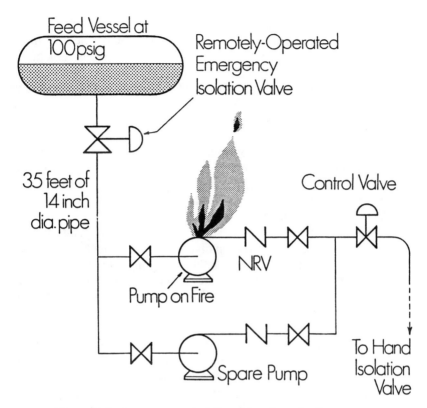

Figure 7.4. An emergency isolation valve stopped a fire.

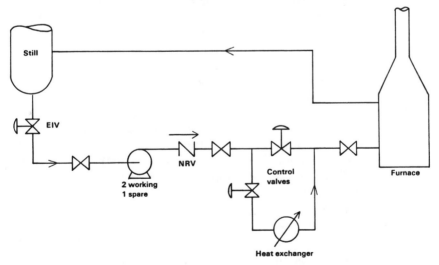

Figure 7.5. Another emergency isolation valve stopped a fire.

The control valve in the delivery line to the furnace was also closed. Unfortunately, this valve was bypassed by the line through the heat exchanger. In the heat of the moment, no one remembered to close the valve in the bypass line. In addition, the nonreturn valve did not hold. The flow of oil backward from the furnace was stopped by closing a hand valve next to the furnace which was about 30 m from the fire. Afterward, another EIV was installed in the pump delivery line.

After the fire, the nonreturn valves on all three pumps were found to be out of order. On one the seat had become unscrewed. On another the fulcrum pin was badly worn. On the third the pin was worn right through and the flap was loose. The valves had not been inspected since the plant was built.

Nonreturn valves have a bad name among many plant operators. However, this is because many of them are never inspected or tested. No equipment, especially containing moving parts, can be expected to work correctly forever without inspection and repair. When nonreturn valves are relied on for emergency isolation they should be scheduled for regular inspection.

The EIV was not affected by the fire. But it was close to it and the incident drew attention to the need to either place EIVs where they are unlikely to be affected by fire or to provide them with fire protection. Fire-resistant sacks are available.

The impulse lines—electrical or pneumatic—leading to EIVs should also be fire-protected.

If control valves are used for emergency isolation, a special switch may be necessary, out in the plant, to close them in an emergency, so that operators do not have to go to the control room to alter the set-points on the controllers.

Note that the operation of an emergency isolation valve should automatically shut down any pump in the line and trip the fuel supply to any furnace.

(c) In contrast, on other occasions, EIVs failed to control fires because the installation was not up to standard. In one case a fire burned for six hours because the button controlling the EIV was too close to the leaking pump for anyone to operate it safely (4). It should have been at least 10 m away. In another case an EIV failed to work because it had not been tested regularly. All EIVs should be tested regularly, say monthly. If they cannot be closed without upsetting production, they should be closed part way and tested fully at shut-downs.

(d) Emergency isolation valves are, of course, of no value if they are not used when required. Sometimes when there has been a leak of a hazardous material the operators have been tempted to try to isolate the leak without shutting down the plant. In doing so they have taken unnecessary risks. For example, there was a bad leak of propylene on a pump inside a building. Four workers were badly injured. Afterward, a lot of money was spent on moving the pumps into the open air, surrounding them with a steam curtain (5) and fitting remotely operated isolation valves and blowdown valves. If another leak should occur, then it would be possible to stop the leak by closing the pump suction valve, opening the blowdown valve and switching off the pump motor without any need for anyone to go near the pumps. See Section 8.1 (c).

Eight years went by before another bad leak occurred. When it did occur the area around the pumps was filled with a visible cloud of propylene vapor 1 m deep. Instead of using the emergency equipment, which would have stopped the flow of propylene and shut down the plant, two very experienced foremen went into the compound, shut down the leaking pump and started the spare up in its place. Fortunately the leak did not fire.

Afterward, when one of the foremen was back in his office, he realized the risk he had been taking. He complained that he should not be expected to take such risks. He had forgotten, in his eagerness to maintain production, that emergency equipment had been provided to avoid the need for such risk taking. Another incident which could have been controlled by an EIV is described in Section 16.1 (g).

7.2.2 OTHER METHODS OF CONTROLLING LEAKS

The following methods have been used successfully:

(a) Injecting water so that it leaks out instead of oil. This method can, of course, be used only when the water pressure is higher than the oil pressure.
(b) Reducing the plant pressure, thus reducing the size of the leak.
(c) Closing an isolation valve some distance away.
(d) Freezing a pipeline. This method requires time to organize the necessary equipment and can only be used with materials of relatively high freezing point such as water or benzene.
(e) Injecting a sealing fluid into a leaking flange or valve gland using a proprietary process such as Furmaniting.
(f) Confining the spread of the leak by water spray (6) or steam curtains (5). The latter have to be permanently installed but the former can be temporary or permanent.

(g) Controlling the evaporation from liquid pools by covering with foam. This method can be used for chlorine and ammonia spillages if suitable foams are used.

7.2.3 HOW NOT TO CONTROL A LEAK

On many occasions employees have entered a cloud of flammable gas or vapor to isolate a leak. In the incident described in Section 7.2.1 (d) this was done to avoid shutting down the plant. More often it has been done because there was no other way of stopping the leak. The persons concerned would have been badly burned if the leak had ignited while they were inside the cloud.

It would be going too far to say that no one should ever enter a cloud of flammable vapor to isolate a leak. There have been occasions when, by taking a risk for a minute, a man has isolated a leak which would otherwise have spread a long way and probably ignited, perhaps exploded. However, we should try to avoid putting people in such situations by providing remotely operated emergency isolation valves to isolate likely sources of leak.

It may be possible to isolate a leak by hand by forcing back the vapor with water spray and protecting the man who closes the valve in the same way. The National Fire Protection Association can provide a set of slides or a film showing how this is done.

It is possible to measure the extent of a leak of flammable gas or vapor with a combustible gas detector. If the leak is small, a person may be allowed (but not expected) to put his hands, suitably protected, inside the flammable cloud. But only in the most exceptional circumstances should a person be allowed to put more of his body into the cloud.

7.3 LEAKS ONTO WATER OR WET GROUND

Section 1.4.4 described two leaks onto pools of water which spread much further than anyone expected. One was ignited by a welder 20 m away and the other spillage, onto a canal, caught fire 1 km away.

In other cases spillages of oil have soaked into the ground and have then come to the surface after heavy rain. A spillage of gasoline in Essex, England, in 1966, came back to the surface two years later. The vapor accumulated on the ground floor of a house, ignited and blew a hole in the stairs, injuring two people. A trench 7 m deep was dug to recover the rest of the rest of the gasoline (7).

In other cases, spillages of oil have leaked into sewers and from there into houses.

If a substantial quantity of oil is spilled into the ground, attempts should be made to recover it by digging a well or trench.

7.4 DETECTION OF LEAKS

On many occasions combustible gas detectors have detected a leak soon after it started and action to control it was taken promptly. Installation of these detectors is strongly recommended whenever liquefied flammable gases or other flashing liquids are handled or experience shows there is a significant chance of a leak (3).

However, these detectors do not do away with the need for regular patrols of the plant by operators. Several plants which have invested heavily in gas detectors report that, nevertheless, half the leaks that have occurred have been detected by people.

REFERENCES

1. *The Explosion and Fire at Chemstar Ltd, 6 September 1981,* Her Majesty's Stationery Office, London, 1982.
2. C. T. Adcock and J. D. Weldon, *Chemical Engineering Progress,* Vol. 63, No. 8, August 1967, p. 54.
3. T. A. Kletz, *Chemical Engineering Progress,* Vol. 71, No. 9, Sept. 1975, p. 63.
4. T. A. Kletz, *Hydrocarbon Processing,* Vol. 58, No. 1, Jan. 1979, p. 243.
5. H. G. Simpson, *Power and Works Engineering,* 8 May 1974, p. 8.
6. J. McQuaid, *Proceedings of the Second International Symposium on Loss Prevention and Safety Promotion in the Process Industries,* Heidelberg, Sept. 1977, Dechema, Frankfurt, p. 511.
7. *Petroleum Times,* 11 April 1969.

Chapter 8

Liquefied Flammable Gases

This Chapter describes a number of incidents involving liquefied flammable gases (LFG) which could have occurred only with these materials (or other flashing flammable liquids).

The property of LFG which makes it so hazardous is that it is usually stored and handled under pressure at temperatures above normal boiling points. Any leak thus flashes, much of it turning to vapor and spray. This can spread for hundreds of meters before it reaches a source of ignition.

The amount of vapor and spray produced can far exceed the theoretical amount of vapor produced, estimated by heat balance (1). The vapor carries some of the liquid with it as spray. It may evaporate on contact with the air. In any case, it is just as likely to burn or explode.

Any flammable liquid under pressure above its normal boiling point will behave like LFG. Liquefied flammable gases are merely the most common example of a flashing liquid. Most unconfined vapor cloud explosions, including the one at Flixborough (Section 2.4), have been due to leaks of such flashing liquids (2).

The term liquefied petroleum gas (LPG) is often used to describe those liquefied flammable gases that are derived from petroleum. The term LFG is preferred. It includes materials such as ethylene oxide, vinyl chloride, and methylamines which behave similarly so far as their flashing and flammable properties are concerned.

LFGs stored at atmospheric pressure and low temperature behave rather differently from those stored under pressure at atmospheric temperature and are not considered in this chapter. An incident involving these materials was described in Section 2.5.

8.1 MAJOR LEAKS

(a) FEYZIN

The bursting of a large pressure vessel at Feyzin, France, in 1966 was one of the worst incidents involving LFG that has ever occurred. It caused many companies to revise their standards for the storage and handling of these materials. Because no detailed account has been published, it is described here. The information is based partly on References 3–6, but also on a discussion with someone who visited the site soon after the fire.

An operator had to drain water from a 1,200 m³ spherical storage vessel nearly full of propane (Figure 8.1). He opened valves A and B. When traces of oil showed that the draining was nearly complete he shut A and then cracked it to complete the draining. No flow came. He opened A fully. The choke—presumably hydrate—cleared suddenly and the operator and the two other men were splashed with liquid. The handle came off valve A and they could not get it back on. Valve B was frozen and could not be moved. Access was poor because the drain valves were immediately below the tank, which was only 1.4 m above the ground.

A visible cloud of vapor, 1 m deep, spread for 150 m and was ignited 25 minutes after the leak started by an automobile that had stopped on a nearby road. The road had been closed by the police but the driver ap-

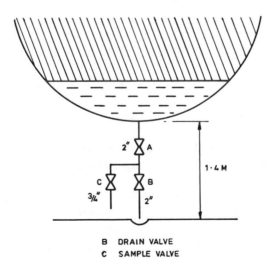

B   DRAIN VALVE
C   SAMPLE VALVE

Figure 8.1. Drain valves underneath propane tank at Feyzin.

proached from a side road. The fire flashed back to the sphere, which was surrounded by flames. There was no explosion. The sphere was fitted with water sprays. But the system was designed to deliver only half the quantity of water normally recommended (0.2 US gallons/ft^2 min or 8 liters/m^2 min) and the supply was inadequate. When the fire brigade started to use their hoses, the supply to the spheres ran dry. The firemen seemed to have used most of the available water for cooling neighboring spheres to stop the fire from spreading, in the belief that the relief valve would protect the vessel on fire.

The ground under the sphere was level so that any propane that did not evaporate or burn immediately collected under the sphere and burned later.

Ninety minutes after the fire started, the sphere burst. Ten out of twelve firemen within 50 m were killed. Men 140 m away were badly burned by a wave of propane which came over the compound wall. Altogether 15–18 men were killed (reports differ) and about 80 injured. The area was abandoned. Flying debris broke the legs of an adjacent sphere which fell over. Its relief valve discharged liquid which added to the fire, and 45 minutes later this sphere burst. Altogether five spheres and two other pressure vessels burst and three were damaged. The fire spread to gasoline and fuel oil tanks.

At first it was thought that the spheres burst because their relief valves were too small. But later it was realized that the metal in the upper portions of the spheres was softened by the heat and lost its strength. Below the liquid level the boiling liquid kept the metal cool. Incidents such as this one in which a vessel bursts because the metal gets too hot are known as Boiling Liquid Expanding Vapor Explosions or BLEVEs.

To prevent such incidents occurring, many companies—after Feyzin—adopted recommendations similar to the following.

RECOMMENDATIONS TO PREVENT A FIRE STARTING

Restrict the size of the second drain valve to 3/4-inch, and place it at least 1 m from the first valve. The drain line should be robust, and firmly supported. Its end should be located outside the shadow of the tank.

Fit a remotely controlled emergency isolation valve (see Section 7.2.1) in the drain line.

New installations should be provided with only one connection below the liquid level, fully welded up to a first remotely operated fire-safe isolation valve located clear of the tank area.

Combustible gas detectors should be installed to provide early warning of a leak.

Further details have been published (7).

RECOMMENDATIONS TO PREVENT A FIRE FROM ESCALATING

Insulate vessels with a fire resistant insulation such as vermiculite concrete. This is available as an immediate barrier to heat input. Unlike water spray, it does not have to be commissioned.

Provide water spray or deluge. If insulation is provided, then water deluge at a rate of 0.06 US gallons/ft² min (2.4 l/m² min) is sufficient. If insulation is not provided, then water spray at a rate of 0.2 US gallons/ft² min (8 l/m² min) is necessary. (Deluge water is poured on the top of a vessel; spray is directed at the entire surface.)

Slope the ground so that any spillage runs off to a collection pit.

Fit an emergency depressuring valve so that the pressure in the vessel can be reduced to one-fifth of design in ten minutes to reduce the strain on the metal. The time can be increased to 30 minutes if the vessel is insulated and to one hour if, in addition, the ground is sloped.

Further details have been published (8). Figure 8.2 summarizes these proposals.

(b) DUQUE DE CAXIAS

A similar incident to that at Feyzin occurred at this refinery in Brazil in 1972. According to press reports, the relief valve failed to open when the pressure in an LPG sphere rose. To try to reduce the pressure, the opera-

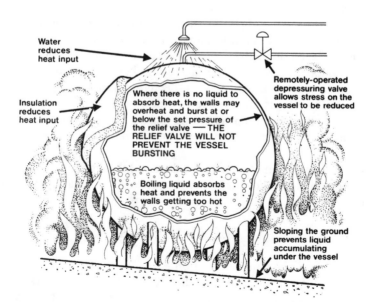

Figure 8.2. Methods of protecting a pressure vessel against fire.

tors opened the drain valve. Little water came out and the LPG that followed it caused the valve to freeze and the flow could not be stopped. There was only one drain valve. The LPG ignited, the vessel BLEVEd and 37 people were killed.

If the operators did, in fact, try to reduce the pressure by draining water, they did not realize that the vapor pressure above a liquid is the same whatever the quantity present.

(c) UNITED KINGDOM

This fire occurred some years ago because those concerned did not fully appreciate the difference in behavior between liquid hydrocarbons, such as naphtha or gasoline, and LFGs. The vapor from a spillage of gasoline will spread only a short distance—about the diameter of the pool. But the vapor from a spillage of LFG can spread for hundreds of meters.

Some equipment which had been designed and used for handling gasoline and similar liquids was adapted to handle propylene. A leak occurred from the gland of a high pressure reciprocating pump operating at 250 bar due to the failure of the studs holding the gland in position. The pump was located in an unventilated building. But the vapor escaped through a large doorway opposite the pump and was ignited by a furnace 75 m away. Four men were badly burned.

The vapor from a spillage of gasoline in the same position would not have spread anywhere near the furnace.

After the fire, the pump (and others) was relocated in the open air, under a canopy, so that small leaks would be dispersed by natural ventilation. It was surrounded by a steam curtain to disperse larger leaks. This would not have been necessary if it could have been located more than 150 m from sources of ignition. Gas detectors were installed to give early warning of any leaks. Emergency isolation valves (Section 7.2.1) were provided so that the pumps could be isolated safely from a distance (9). What happened when another leak occurred was described in Section 7.2.1 (d).

Note that a common factor in incidents (a) through (c) was a failure by those concerned to understand the properties of the materials and equipment.

8.2 MINOR LEAKS

(a) Following Feyzin (see Section 8.1 (a)) one company spent a lot of money improving the standard of its LFG storage facilities—in particular, the water-draining arrangements so as to comply with the recommendations made in Section 8.1 (a).

Less than a year later a small leak from a passing drain valve on a pipeline caught fire. It was soon extinguished by closing the valve. But an investigation disclosed that:

1. There should have been two valves in series or a single valve and blank.
2. The valve was made of brass and was of a type stocked for use on domestic water systems. It was not the correct pressure rating for LFG.
3. The valve was screwed onto the pipeline, though the company's codes made it clear that only flanged or welded joints were allowed.
4. It was never discovered who installed this unauthorized substandard drain point. An attempt had been made to publicize the lessons of Feyzin, the company's standards and the reasons for them. However, this did not prevent the installation of the drain point. Note that a number of people must have been involved. As well as the man who actually fitted it and his foreman, someone must have issued a work permit and accepted it back (when he should have inspected the job) and several persons must have used the drain point. Many must have passed by. If only one of them had recognized the substandard construction and drawn it to the attention of those responsible, the fire would not have occurred (10).

Like the plants in our gardens, our plants grow unwanted branches while our backs are turned.

(b) A propane sphere was filled with water to inert it during repair work. When the repairs were complete, the water was drained from the sphere and propane vapor was admitted to the top to replace the water. The instruction stated that draining should stop when 5 m³ of water was left in the vessel. But there was no one present when this stage was reached. All the water drained out, followed by propane. Fortunately it did not ignite. The job had been left because the operators did not realize that the level indicator, which measured weight, would indicate a level of water almost twice the actual level. Other similar incidents are described in Section 5.1.2. If nitrogen is available, it should be used instead of water for inerting vessels. Or if water is used, it should be replaced by nitrogen when it is drained. Before filling any equipment with water, always check that it is strong enough to take the weight of the water (11).

8.3 OTHER LEAKS

Numerous leaks of LFG, mainly minor but occasionally more serious, have occurred from the following items of equipment:

FLANGED JOINTS
The size and frequency of leaks can be reduced by using spiral-wound gaskets in place of compressed asbestos fiber ones. Screwed joints should not be used.

PUMP SEALS
The leak size can be reduced by using double mechanical seals or a mechanical seal and a throttle bush, the space between the two being vented to a safe place. Major leaks may still occur, however, due to collapse of the bearing or seal. LFG pumps should therefore be fitted with emergency isolation valves (see Section 7.2.1), particularly if the temperature is low or the inventory that can leak out is high.

LEVEL GLASSES
These should not be used with flashing flammable liquids. See Section 7.1.4.

SAMPLE POINTS
These should not exceed 1/4-inch diameter.

SMALL BRANCHES
These should be physically robust and well-supported so that they cannot be knocked off accidentally or vibrate until they fail by fatigue.

EQUIPMENT MADE FROM GRADES OF STEEL UNSUITABLE FOR USE AT LOW TEMPERATURES
Materials of construction should be chosen so that the equipment will withstand the lowest temperature that can be reached during abnormal operation. In the past materials have been used which will withstand normal operating temperatures but which may become brittle at lower temperatures reached during plant upsets or abnormal operation. Some spectacular failures have resulted (12).

Wholesale replacement of such materials in existing plants is impractical and there is no universal solution. Some lines can be replaced in different grades of steel. Sometimes low temperature trips or alarms can be used. Sometimes the need to watch the temperature closely during startup has to be impressed on operators.

Other leaks of LFG are described in Sections 1.5.4 (a) and (b), 9.1.6 (d) and (f), and 13.4.

REFERENCES

1. J. D. Reed, *Proceedings of the First International Symposium on Loss Prevention and Safety Promotion in the Process Industries,* Elsevier, Amsterdam, 1977, p. 191.
2. J. A. Davenport, *Chemical Engineering Progress,* Vol. 73, No. 9, Sept. 1977, p. 54.
3. *The Engineer,* 25 March 1966, p. 475.
4. *Paris Match,* No. 875, 15 Jan. 1966.
5. *Fire,* special supplement, Feb. 1966.
6. *Petroleum Times,* 21 Jan. 1966, p. 132.
7. Imperial Chemical Industries Ltd, *Liquefied Flammable Gases— Storage and Handling,* Royal Society for the Prevention of Accidents, Birmingham, 1970.
8. T. A. Kletz, *Hydrocarbon Processing,* Vol. 56, No. 8, August 1977, p. 98.
9. T. A. Kletz, *Loss Prevention,* Vol. 13, 1980, p. 1.
10. T. A. Kletz, *Chemical Engineering Progress,* Vol. 72, No. 11, Nov. 1976, p. 48.
11. *Petroleum Review,* April 1982, p. 35.
12. A.L.M. van Eijnatten, *Chemical Engineering Progress,* Vol. 73, Sept. 1977.

Chapter 9

Pipe and Vessel Failures

9.1 PIPE FAILURES

Davenport (1) has listed over 60 major leaks of flammable materials, most of which resulted in serious fires or unconfined vapor cloud explosions. Table 9.1, derived from his data, classifies the leak by point of origin and shows that pipe failures accounted for half the failures—over half if we exclude transport containers.

It is therefore important to know why pipe failures occur. Following, a number of typical failures (or near failures) are discussed.

These and other failures, summarized in References 2 and 3, show that by far the biggest single cause of pipe failures has been the failure of construction teams to follow instructions or to do well what was left to their discretion. The most effective way of reducing pipe failures is to:

1. Specify designs in detail.
2. Check construction closely to see that the design has been followed and that details not specified have been constructed according to good engineering practice.

Many publications on pipe failures attribute them to causes such as fatigue or inadequate flexibility. This is not very helpful. It is like saying a fall was caused by gravity. We need to know why fatigue occurred or why the flexibility was inadequate. To prevent further incidents, should we improve the design, construction, operations, maintenance, inspection, or what? The following incidents, and many others, suggest that improvement should be made to the design/construction interface. That is, we should focus on the detailing of the design, see that it has been followed, and that good practice has been followed when details are not specified.

9.1.1 DEAD-ENDS

Dead-ends have caused many pipe failures. Water, present in traces in many oil streams, collects in dead-ends and freezes, breaking the pipe. Or corrosive materials dissolve in the water and corrode the line.

For example, there was a dead-end branch 12 inches in diameter, 3 m long, in a natural gas pipeline operating at a gauge pressure of 550 psi (38 bar). Water and impurities collected in the dead-end which corroded and failed. The escaping gas ignited at once, killing three men who were looking for a leak (4).

There are other sorts of dead-ends besides pipes which have been blanked. Valved branches which are rarely used are just as dangerous. The feed line to a furnace (Figure 9.1) was provided with a permanent steam connection for use during de-coking.

The connection was on the bottom of the feedline and the steam valve was not close to the feedline. Water collected above the steam valve, froze during cold weather and ruptured the line, allowing oil at a gauge pressure of 450 psi (30 bar) to escape.

If dead-ends cannot be avoided, they should be connected to the *top* of the main pipeline.

An unusual—and unnecessary—dead-end was a length of 2-inch pipe welded onto a process line to provide a support for an instrument (Figure 9.2).

Water collected in the support. Four years after it had been installed, the process line corroded through and a leak of liquefied gas occurred.

Table 9-1
Origin of Leaks Causing Vapor Cloud Explosions

Origin of leak	Number of incidents	Notes
Transport container	10	includes 1 zeppelin
Pipeline (incl. valve, flange, sight-glass etc)	34	includes 1 sight-glass and 2 hoses
Pump	2	
Vessel	5	includes 1 internal explosion, 1 slop-over and 1 failure due to overheating
Relief valve or vent	8	
Drain valve	4	
Error during maintenance	2	
Unknown	2	
Total	67	

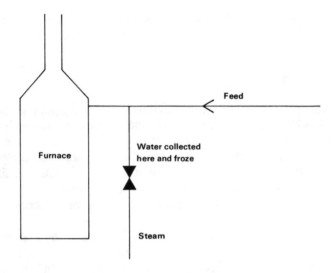

Figure 9.1. The steam connection to the furnance formed a dead-end.

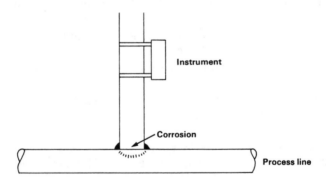

Figure 9.2. Water collected in the instrument support and corroded the process line.

Another serious failure occurred because water in a dead-end was suddenly vaporized. A heavy oil was dried by heating it to 120°C in a tank filled with steam coils. The oil was circulated while it was being dried. The suction line projected into the conical base of the tank, forming a dead-end, as shown in Figure 9.3.

As long as the circulation pump was kept running, water could not settle out in the dead-end. The foreman knew that the pump had to be kept running. When he was transferred to another plant, this information was lost and the pump was used only for emptying the tank.

This worked satisfactorily for a time until some water collected in the dead-end and gradually warmed up as the oil was heated. When the temperature reached 100°C the water vaporized with explosive violence and burst the equipment. The escaping oil caught fire, five men were killed, and the tank ended up in the plant next door.

This incident illustrates the dangers of dead-ends and the pressures developed when water is suddenly vaporized. It also shows how easily knowledge can be lost when people leave. Even if the new foreman was *told* to run the pump all the time or if this was written in the instructions, the reason for doing so might be forgotten and the circulation might be stopped as unnecessary or to save electricity.

Other incidents caused by the sudden vaporization of water are described in Sections 12.2 and 12.4.5.

9.1.2 POOR SUPPORT

Pipes have often failed because their support was insufficient, and they were free to vibrate. On other occasions they failed because their support was too rigid and they were not free to expand.

(a) Many small-bore pipes have failed by fatigue because they were free to vibrate. Supports for these pipes are usually run on site and it is not apparent until startup that the supports are inadequate. It is very easy for the startup team, busy with other matters, to ignore the vibrating pipes until they have more time to attend to them. Then they get so used to them that they do not notice them.

(b) A near failure of a pipe is illustrated in Figure 9.4. An expansion bend on a high-temperature line was provided with a temporary

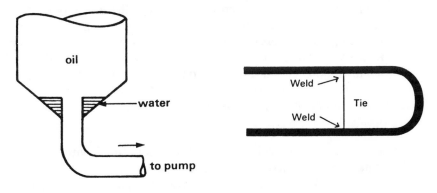

Figure 9.3. Water in this dead-end was vaporized by oil.

Figure 9.4. A construction support on an expansion bend was left in position.

support to make construction easier. It was then left in position. Fortunately, while the plant was coming onstream someone noticed it and asked what it was for.

(c) After a crack developed in a 22-inch-diameter steam main, operating at a gauge pressure of 250 psi (17 bar) and a temperature of 365°C, the main was checked against the design drawings. Many of the supports were faulty. For example, on four successive supports:

1. On No.1 the spring was fully compressed.
2. No. 2 was not fitted.
3. No.3 was in position but not attached to the pipe.
4. No.4 was attached but the nuts on the end of the support rod were slack.

Piping 12-inch-diameter and over is usually tailored for the particular duty. There is a smaller factor of safety than with smaller sizes. With these large pipes it is even more important than with smaller ones that the finished pipework is closely inspected to confirm that the construction team has followed the designer's instructions.

(d) A pipe was welded to a steel support which was bolted to a concrete pier. A second similar support was located 2 m away. The pipe survived normal operating conditions. But when it got exceptionally hot a segment of the pipe was torn out. The fracture extended almost completely around the weld. The bolts anchoring the support to the concrete pier were bent.

This incident was reported in the safety bulletin of another company. The staff dismissed the incident. "Our design procedures," they said, "would prevent it happening." A little later it did happen. A reflux line was fixed rigidly to brackets welded to the shell of a distillation column. At startup the differential expansion of the hot column and the cold line tore one of the brackets from the column. Flammable vapor leaked out but fortunately did not catch fire.

(e) A 10-inch pipe carrying oil at 300°C was fitted with a ³/₄-inch branch on its underside. The branch was located five inches from a girder on which the pipe rested. When the pipe was brought into use, the expansion was sufficient to bring the branch into contact with the girder and knock it off. Calculations showed that the branch would move over six inches.

9.1.3 WATER INJECTION

Water was injected into an oil stream using the simple arrangement shown in Figure 9.5. Corrosion occurred near the point shown and the oil leak caught fire (5). The rate of corrosion far exceeded the corrosion allowance of 0.05 inch per year.

A better arrangement is shown in Figure 9.6. The dimensions are chosen so that the water injection pipe can be removed for inspection.

However, this system is not foolproof. One system of this design was assembled with the injection pipe pointing upstream instead of downstream. This increased corrosion.

As discussed in Section 3.2.1, equipment should be designed so that it is difficult or impossible to assemble it incorrectly or so that the incorrect assembly is immediately apparent.

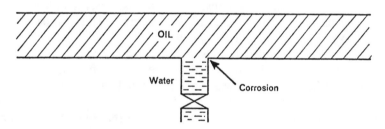

Figure 9.5. Water injection—a poor arrangement.

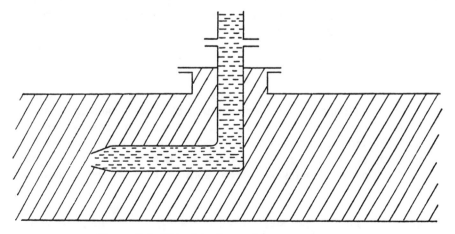

Figure 9.6. Water injection—a better arrangement.

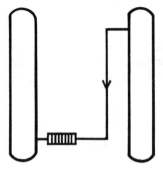

Figure 9.7. A large bellows between the two halves of a distillation column.

9.1.4 BELLOWS

Bellows are a good example of equipment which is intolerant of poor installation or departure from design conditions. They should therefore be avoided on lines carrying hazardous materials. This can be done by building expansion loops into the pipelines.

The most spectacular bellows failure of all time (Flixborough) was described in Section 2.4. Figure 9.7 illustrates a near failure.

A large distillation column was made in two halves, connected by a 42-inch vapor line containing a bellows. During a shutdown this line was steamed. Immediately afterward it was noticed that one end of the bellows was seven inches higher than the other although it was designed for a maximum difference of three inches. It was then found that the design contractor had designed the line for normal operation. But, he had not considered conditions that might be developed during abnormal procedures such as startup and shutdown.

9.1.5 WATER HAMMER

Water hammer occurs in two distinct ways: when the flow of liquid in a pipeline is suddenly stopped, for example, by quickly closing a valve, and when slugs of liquid in a gas line are set into motion by movement of the gas. The latter occurs when condensate is allowed to accumulate in a steam main, because the traps are too few, or out of order or in the wrong place. High pressure mains have been ruptured, as in the following incident.

(a) A 10-inch-diameter steam main operating at a gauge pressure of 600 psi (40 bar) suddenly ruptured, injuring several workers.

The incident occurred soon after the main had been brought back into use following a turnaround. It was up to pressure but there was no flow along it. The steam trap was leaking and had been isolated. An attempt was made to get rid of condensate through the bypass valve. But steam entered the condensate header and the line was isolated, as shown in Figure 9.8. Condensate then accumulated in the steam main.

When a flow was started along the steam main by opening a 3/4-inch valve leading to a consuming unit, the movement of the condensate fractured the main (6).

(b) Figure 9.9 shows how another steam main—this time one operating at a gauge pressure of 20 psi (1.4 bar)—was burst by water hammer.

Two drain points were choked and one isolated. In addition, the change in diameter of the main provided an opportunity for condensate to accumulate. The main should have been constructed so that the bottom was straight and the change in diameter took place at the top.

9.1.6 MISCELLANEOUS PIPE FAILURES

(a) Many failures have occurred because old pipes were reused. For example, a hole six inches long and two inches wide appeared on a three-inch pipe carrying flammable gas under pressure. The pipe

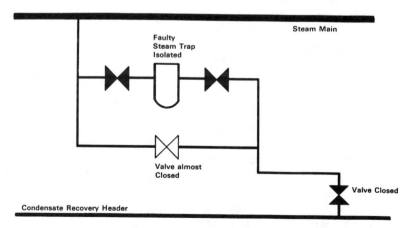

Figure 9.8. Arrangement of valves on steam main which was broken by water hammer.

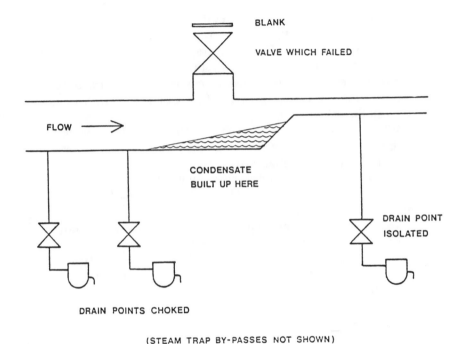

Figure 9.9. Arrangement of drains on steam main which was broken by water hammer.

had previously been used on a corrosive/erosive duty and its condition was not checked before reuse.

In another case a 4½-inch-diameter pipe carrying a mixture of hydrogen and hydrocarbons at a gauge pressure of 3,600 psi (250 bar) and a temperature of 350 to 400°C burst, producing a jet of flame over 30 m long (Figure 9.10). Fortunately, the pipe was located high up and no one was injured.

The grade of steel used should have been satisfactory for the operating conditions. Investigation showed, however, that the pipe had previously been used on another plant for 12 years at 500°C. It had used up a lot of its creep life.

Old pipes should never be reused unless their history is known in detail and tests show they are suitable.

(b) Many failures have occurred because the wrong grade of steel was used for a pipeline. The correct grade is usually specified but the wrong grade is delivered to site or selected from the pipe store.

Figure 9.10. An old pipe was reused and failed by creep.

The most spectacular failure of this sort occurred when the exit pipe from a high-pressure ammonia converter was constructed from carbon steel instead of $1\frac{1}{4}$% Cr, 0.5% Mo. Hydrogen attack occurred and a hole appeared at a bend. The hydrogen leaked out and the reaction forces pushed the converter over.

Many companies now insist that if use of the wrong grade of steel can affect the integrity of the plant, all steel must be checked for composition before use. This applies to flanges, bolts, welding rods, etc., as well as the raw pipe. Steel can be analyzed easily with a spectrographic analyzer. Other failures caused by the use of the wrong construction material are described in Section 16.1.

(c) Several pipe failures have occurred because reinforcement pads have been welded to pipe walls, to strengthen them near a support or branch, and the spaces between the pads and the walls were not vented. For example, a flare main collapsed, fortunately while it was being stress relieved.

Pipe reinforcement pads can be vented by intermittent rather than continuous welding or a $1/8$-inch or $1/4$-inch hole can be drilled in the pad.

(d) Corrosion—internal or external—often causes "leak before break" failures, but not always.

A line carrying liquefied butene at a gauge pressure of about 30 psi (2 bar) passed through a pit where some valves were located. The pit was full of water, contaminated with some acid. The pipe corroded and a small leak occurred. The line was emptied for repair by flushing with water at a gauge pressure of 110 psi (7.5 bar). The line was designed to withstand this pressure. However, in its corroded state it could not do so and the bonnet was blown off a valve. The operator isolated the water. This allowed butene to flow out of the hole in the pipe. Twenty minutes later the butene exploded, causing extensive damage (7).

(e) A 1-inch screwed nipple and valve blew out of an oil line operating at 350°C. The plant was covered by an oil mist which ignited 15 minutes later. The nipple had been installed about 20 years earlier, during construction, to facilitate pressure testing. It was not shown on any drawing and its existence was not known to the operating team. If they had known it was there they would have replaced it with a welded plug.

Similar incidents are described in Section 7.1.5.

(f) Not all pipe failures are due to inadequacies in design or construction (for example, the one described in Section 1.5.2).

A near failure was also due to poor maintenance practice. A portable hand-held compressed air grinder was left resting in the space between two live lines. The switch had been left in the "on" position. So when the air compressor was started the grinder started to turn. It ground away part of a line carrying liquefied gases. Fortunately the grinder was noticed and removed before it had ground right through the line, but it reduced the wall thickness from 0.28 inch to 0.21 inch.

9.1.7 FLANGE LEAKS

Leaks from flanges are more common than those described in Sections 9.1.1–9.1.6, but are also usually smaller. On lines carrying LFGs and other flashing liquids, spiral-wound gaskets should be used in place of compressed asbestos fiber (caf) gaskets because they restrict the size of any leak to a very small value. A section of a caf gasket between two bolts has often blown out, causing a fair-sized leak. But this will not occur with a spiral-wound gasket.

9.2 PRESSURE VESSEL FAILURES

Failures of pressure vessels are very rare. Many of those that have been reported occurred during pressure test or were cracks detected during routine examination. Major failures leading to serious leaks are hard to find.

Low-pressure storage tanks are much more fragile than pressure vessels. They are therefore more easily damaged. Some failures were described in Chapter 5.

A few vessel failures and near-failures are described next—to show that they can occur. Failures of vessels as a result of exposure to fire were described in Section 8.1.

9.2.1 FAILURES PREVENTABLE BY BETTER DESIGN OR CONSTRUCTION

These are hard to find.

(a) A leak of gas occurred through the weep hole in a multiwall vessel in an ammonia plant. The plant stayed on line but the leak was watched to see that it did not worsen. Ten days later the vessel disintegrated, causing extensive damage.

The multiwall vessel was made from an inner shell and 11 layers of wrapping, each drilled with a weep hole. The disintegration was attributed to excessive stresses near a nozzle. These had not been recognized when the vessel was designed.

The report on the incident states: "Our reading of the literature led us to believe that as long as the leaking gas could be relieved through the weep holes, it would be safe to operate the equipment. We called a number of knowledgeable people and discussed the safety issue with them. Consensus at the time supported our conclusion. But after the explosion, there was some dispute over exactly what was said and what was meant. Knowing what we know now, there can be no other course in the future than to shut down operations in the event of a leak from a weep hole under similar circumstances." (8).

(b) An ammonia plant vessel disintegrated as the result of low cycle fatigue—the result of repeated temperature and pressure cycles (9).

(c) An internal ball float in a propane storage sphere came loose. When the tank was overfilled, the ball lodged in the short pipe leading to the relief valve, in which it formed an exact fit. When the sphere warmed up, the rise in pressure caused its diameter (14 m) to increase by 0.15 m (6 inches). The increase in diameter was noticed when it was found that the access stairway had broken loose.

If ball-floats are used, their dimension and those of the vessel should be checked. If a similar incident could occur, then the relief pipe should be protected by a cage.

9.2.2 FAILURES PREVENTABLE BY BETTER OPERATION

The incident described in Section 9.2.1 (a) might be classified in this way.

(a) Low-pressure storage tanks have often been sucked in, as described in Section 5.3. Pressure vessels can also be sucked in if they have not been designed to withstand vacuum, as the following incident shows.

A blowdown drum was taken out of service and isolated. The drain line was removed and a steam lance inserted to sweeten the tank. The condensate ran out of the same opening.

The condensate was isolated and 45 minutes later the drain valve was closed. Fifteen minutes later the vessel collapsed.

Clearly, 45 minutes was not long enough for all the steam to condense.

(b) A redundant pressure vessel, intended for reuse at atmospheric pressure, had been installed by contractors who decided to pressure test it. They could not find a water hose to match any of the connections on the vessel. They therefore decided to pressure test it with compressed air. The vessel reached a gauge pressure of 25 psi (1.7 bar) before it ruptured.

It is possible that the employees concerned did not understand the difference between a pressure test, normally carried out with water, and a leak test, often carried out with compressed air at a pressure well below the test pressure.

This incident shows the need to define the limits within which contractors can work and to explain these limits to contractors' employees.

Another incident in which a pressure vessel was ruptured by compressed air, this time because the vent was choked, was described in Section 2.2 (a).

(c) A vessel, designed to operate at a gauge pressure of 5 psi (0.3 bar) and protected by a rupture disc, was being emptied by pressurization with compressed air. The operator was told to keep the gauge pressure below five psi but he did not do so and the vessel burst, spraying him with a corrosive chemical. A valve below the rupture disc was closed and had probably been closed for some time.

It is bad practice (and in some countries illegal) to fit a valve between a vessel and its rupture disc (or relief valve). The valve had been fitted to stop escapes of gas into the plant after the disc had blown and while it was being replaced. A better way, if isolation is required, is to fit two rupture discs, each with its own isolation valve, the valves being interlocked so that one is always open.

If compressed gas has to be blown into a vessel which cannot withstand its full pressure, then it is good practice to fit a reducing valve on the gas supply. This would be possible in the case just described. But it may not be possible if the gas is used to blow liquid *into* a vessel. If the gas pressure is restricted to the design pressure of the vessel, it may not be sufficient to overcome friction and change in height.

A sidelight on the incident is that the operator had worked on the plant for only seven months and during that time had received five warnings for lack of attention to safety or plant operations. However, the incident was not due to the operator's lack of attention, it was due to the poor design of the equipment. Sooner or later, a valve will be shut when it should be open or vice versa and the design or method of operation should allow for this. See also Section 1.1 on isolation for maintenance.

REFERENCES

1. J. A. Davenport, *Chemical Engineering Progress,* Vol. 73, No. 9, Sept, 1977, p. 54.
2. T. A. Kletz, *Proceedings of the Fourth International Conference on Pressure Vessel Technology,* Institution of Mechanical Engineers, London, 1980, p. 25.
3. T. A. Kletz, *Plant/Operation Progress,* Vol. 3, No. 1, Jan. 1984, p. 19.
4. U.S. National Transportation Safety Board, *Safety Recommendations,* P-75-14 & 15, 1975.
5. *The Bulletin, The Journal of the Association for Petroleum Acts Administration,* April 1971.
6. *Explosion from a steam line: Report of Preliminary Inquiry No. 3471,* Her Majesty's Stationery Office, London, 1975.
7. C. H. Vervalin, *Fire Protection Manual for Hydrocarbon Processing Plants,* Vol. 1, 3rd Edition, Gulf Publishing Co., 1985, p. 122.
8. L. B. Patterson, *Ammonia Plant Safety,* Vol. 21, 1979, p. 95.
9. J. E. Hare, *Plant/Operations Progress,* Vol. 1, No. 3, July 1982, p. 166.

Chapter 10

Other Equipment

Incidents involving storage tanks, stacks, pipelines and pressure vessels have been described in Chapters 5, 6, and 9. This chapter describes some incidents involving other items of equipment.

10.1 CENTRIFUGES

Many explosions, some serious, have occurred in centrifuges handling flammable solvents because the nitrogen blanketing was not effective.

In one case a cover plate between the body of the centrifuge and a drive housing was left off. The nitrogen flow was not large enough to prevent air entering and an explosion occurred, killing two men. The source of ignition was probably sparking caused by the drive pulley which had slipped and fouled the casing. However, the actual source of ignition is unimportant. In equipment such as centrifuges which contains moving parts, sources of ignition can easily arise.

In another incident the nitrogen flow was too small. The range of the rotameter in the nitrogen line was 0–60 l/min (0–2 ft³/min) although 150 l/min (5 ft³/min) was needed to keep the oxygen content at a safe level.

On all centrifuges which handle flammable solvent the oxygen content should be continuously monitored. At the very least, it should be checked every shift with a portable analyzer. In addition, the flow of nitrogen should be adequate, clearly visible and read regularly.

These recommendations apply to all equipment which is blanketed with nitrogen, including tanks (Section 5.4) and stacks (Section 6.1). But they are particularly important for centrifuges due to the ease with which sources of ignition can arise (1).

Another hazard with centrifuges is that if they turn the wrong way the snubber can damage the basket. It is therefore much more important than with pumps to make sure this does not occur.

10.2 PUMPS

The biggest hazard with pumps is failure of the gland, sometimes the result of bearing failure, leading to a massive leak of flammable, toxic or corrosive chemicals. Often it is not possible to get close enough to the pump suction and delivery valves to close them. Many companies therefore install remotely operated emergency isolation valves in the suction lines (and sometimes in the delivery lines as well) as discussed in Section 7.2.1. A nonreturn valve (check valve) in the delivery line can be used instead of an emergency isolation valve, provided it is scheduled for regular inspection.

Another common cause of accidents with pumps is dead-heading—that is, allowing the pump to run against a closed delivery valve. This has caused rises in temperature, leading to damage to the seals and consequent leaks. It has caused explosions when the material in the pump decomposed at high temperature. In one incident air, saturated with oil vapor, was trapped in the delivery pipework. Compression of this air caused its temperature to rise above the auto-ignition temperature of the liquid and an explosion occurred—a diesel engine effect.

Positive pumps are normally fitted with relief valves. These are not usually fitted to centrifugal pumps unless the process material is likely to explode if it gets too hot. As an alternative to a relief valve, such pumps may be fitted with a high-temperature trip. This isolates the power supply. Or a kick-back, a small-bore line (or a line with a restriction orifice plate) leading from the delivery line back to the suction vessel may be used. The line or orifice plate is sized so that it will pass just enough liquid to prevent the pump from overheating.

Pumps fitted with an auto-start will dead-head if they start when they should not. This has caused overheating. Such pumps should be fitted with a relief valve or one of the other devices just described.

In another incident a condensate pump was started up by remote operation with both suction and delivery valves closed. The pump disintegrated, bits being scattered over a radius of 20 m.

If remote starting must be used, then some form of interlock is needed to prevent similar incidents from occurring.

Pumps can overheat if they run with the delivery valve almost closed. In one incident a pump designed to deliver 10 t/h was required to deliver only ¼ t/h. The delivery valve was gagged, the pump got too hot, the casing joint sprang and the contents leaked out and caught fire.

If a pump is required to deliver a very small fraction of its design rate, a kick-back should be provided.

Many bearing failures and leaks have occurred as the result of lack of lubrication. Sometimes operators have neglected to lubricate the pumps. On one occasion, a bearing failure was traced to water in the lubricating oil. The bearing failure caused sparks which set fire to some oily residues nearby. Drums of oil in the open-air lubricating oil storage area were found to be open so that rain water could get in. This is a good example of high-technology—in bearing and seal design—frustrated by a failure to attend to simple things.

More than any other item of equipment, pumps require maintenance while the rest of the plant is on line. Many incidents have occurred because pumps under repair were not properly isolated from the running plant. See Section 1.1.

10.3 AIR COOLERS

A pump leak caught fire. There was a bank of fin-fan coolers above the pump and the updraft caused serious damage to the coolers. There was an emergency isolation valve in the pump suction line. This was soon closed and the fire extinguished, but not before the fin-fans were damaged. The damage to them was far greater than the damage to the pumps. It is not good practice to locate fin-fans over pumps or other equipment which is liable to leak.

On several other occasions the draft from fin-fans has made fires worse. And the fans could not be stopped because the stop buttons were too near the fire. The stop buttons should be located (or duplicated) at least 10 m away.

Another hazard with air coolers is that even though the motor is isolated, air currents have caused the fans to rotate while they were being maintained. They should therefore be prevented from moving before any maintenance work is carried out on or near them.

10.4 RELIEF VALVES

Very few incidents occur because of faults in relief valves themselves. When equipment is damaged because the pressure could not be relieved it is usually found afterwards that the relief valve (or other relief device) had been isolated [see Section 9.2.2 (c)], wrongly installed [see Section 3.2.1 (e)] or interfered with in some other way [see Section 9.2.1 (c)]. The following incidents are concerned with the peripherals of relief valves rather than the valves themselves.

10.4.1 LOCATION

A furnace was protected by a relief valve on its inlet line (Figure 10.1).

A restriction developed after the furnace. The relief valve lifted and took most of the flow. The flow through the furnace tubes fell to such a low level that they overheated and burst. The low-flow trip, which should have isolated the fuel supply to the furnace when the flow fell to a low value, could not do so because the flow through it was normal.

The relief valve should have been placed after the furnace or, if this was not possible, before the low-flow trip.

Another point on location, which is sometimes overlooked, is that most relief valves are designed to be mounted vertically and should not be mounted horizontally.

10.4.2 RELIEF VALVE REGISTERS

All companies keep a register of relief valves. They test them at regular intervals (every one or two years) and do not allow their sizes to be changed without proper calculation and authorization.

However, equipment has been overpressured because the following items were not registered. They had been overlooked because they were not obviously a relief device or part of the relief system.

(a) A hole or an open vent pipe—the simplest relief device possible. Section 2.2 (a) described how two men were killed because the size of a vent hole in a vessel was reduced from six inches to three inches.

(b) A restriction orifice plate limiting the flow into a vessel or the heat input into a vessel should be registered if it was taken into account in sizing the vessel's relief valve.

Restriction plates are easily removed. A short length of narrow bore pipe is better.

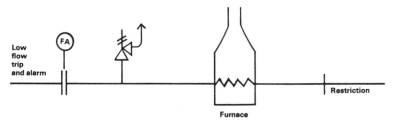

Figure 10.1. When the relief valve lifted, the flow through the furnace was reduced.

(c) A control valve limiting the flow into a vessel or the heat input into a vessel should be registered if its size was taken into account in sizing the vessel's relief valve. The control valve record sheets should be marked to show that the trim size should not be changed without checking that the relief valve will still be suitable.

(d) Nonreturn valves (check valves) should be registered *and inspected regularly* if their failure could cause a relief valve to be undersized. Usually two nonreturn valves *of different types* in series are used if the nonreturn valve forms part of the relief system.

10.4.3 CHANGING RELIEF VALVES

Some vessels are provided with two full-size relief valves so that one can be changed with the plant online. On the plant side of the relief valves isolation valves are usually provided below each relief valve, interlocked so that one relief valve is always open to the plant (Figure 10.2). If the relief valves discharge into a flare system, it is not usual to provide such valves on the flare side. Instead the relief valve is simply removed and a blank fitted quickly over the end of the flare header, before enough air is sucked in to cause an explosion. Later the blank is removed and the relief valve replaced.

On one plant a fitter removed a relief valve and then went for lunch before fitting the blank. He returned just as an explosion occurred. He

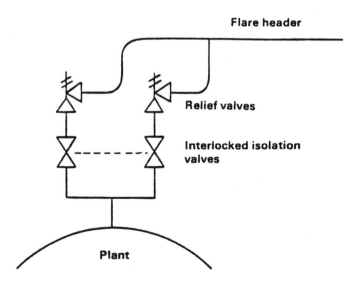

Figure 10.2. Two relief valves with interlocked isolation valves.

was not injured by the explosion but was slightly injured sliding down a pipe to escape quickly.

Removing a valve and fitting a blank is satisfactory if the operators make sure, before the relief valve is removed, that the plant is steady and that this relief valve or any other is unlikely to lift. Unfortunately such instructions may lapse with the passage of time. This occurred at one plant. They were fully aware that air might get into the flare system. They knew about the incident just described. But they were less aware that oil might get out. While an 8-inch relief valve was being changed, another relief valve lifted and gasoline came out of the open end. Fortunately it did not ignite.

The investigation showed that at the time the operating team was busy at the main plant which they operated. Changing the relief valve on the auxiliary unit had been left in charge of a deputy foreman. He wanted to get it done while a crane was available.

The best way to change a relief valve when a plant is on line is to use the sealing plate shown in Figure 10.3.

All but two of the bolts joining the relief valve to the flare header are removed. The sealing plate is then inserted between the relief valve and the flare header. It is secured by special bolts with small heads that pass through the bolt holes in the relief valve flange, but not through the holes in the sealing plate. The last two bolts can then be removed and the relief valve removed. To replace the relief valve the procedure is reversed (2).

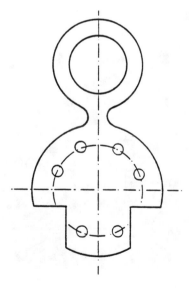

Figure 10.3. Sealing plate for changing relief valves.

This system is recommended for changing relief valves on lines greater than four inches diameter.

Flare lines should never be slip-plated with ordinary slip-plates because they may be left in position in error. The sealing plate cannot be left in position when the relief valve is replaced.

When replacing relief valves, care is needed that the right relief valve is replaced. Valves of different internal size many look alike. See Section 1.2.4.

10.4.4 TAIL PIPES

Figure 10.4 shows what happened to the tailpipe of a steam relief valve that was not adequately supported. The tailpipe was not provided with a drain hole (or if one was provided it was too small) and the tailpipe filled with water. When the relief valve lifted, the water hit the curved top of the tailpipe with great force.

On other occasions drain holes have been fitted in relief valve tailpipes even though the relief valve discharged into a flare system. Gas has then escaped into the plant area.

Figure 10.4. This relief valve tailpipe was not adequately supported.

On occasions, relief valve exit pipes that have not been adequately supported and have sagged on exposure to fire, restricting the relief valve discharge.

10.4.5 RELIEF VALVE FAULTS

Finally here are a few examples of faults in relief valves themselves. These are not the result of errors in design but of poor maintenance practice. The following have all been seen:

1. Identification numbers stamped on springs thus weakening them.
2. The sides of springs ground down so that they fit.
3. Corroded springs.
4. A small spring put inside a corroded spring to maintain its strength. Sometimes the second spring was wound the same way as the first spring so that the two interlocked.
5. Use of washers to maintain spring strength.
6. Welding of springs to end caps.

Do not assume that such things could not happen in your company (unless you have spent some time in the relief valve workshop).

REFERENCES

1. *Users guide for the safe operation of centrifuges with particular reference to hazardous atmosphere,* Institution of Chemical Engineers, 1976.
2. T. A. Kletz, Loss Prevention, Vol. 6, 1972, p. 134.

Chapter 11

Entry to Vessels

Many people have been killed or overcome because they entered vessels or other confined spaces which had not been thoroughly cleaned or tested. A number of incidents are described here. Others involving nitrogen are described in Section 12.3. Sometimes it seems that vessels are more dangerous empty than full.

For further details of the procedures that should be followed when preparing vessels for entry, see References 1 and 2.

11.1 VESSELS NOT FREED FROM HAZARDOUS MATERIAL

In these incidents the vessels were correctly isolated but were not freed from hazardous materials.

(a) A vessel was divided into two halves by a baffle which had to be removed. The vessel was cleaned out, inspected and a permit issued for a worker to enter the left-hand side of the vessel to burn out the baffle. It was impossible to see into the right-hand half. But because the left-hand half was clean and because no combustible gas could be detected it was assumed that the other half was also clean (Figure 11.1). While the welder was in the vessel some deposit in the right-hand half caught fire. The welder got out without serious injury but bruised himself in his haste.

If a part of the vessel cannot be inspected and be seen to be safe, then we should assume the vessel contains hazardous materials.

If the previous contents were flammable, we should assume there is some flammable material out of sight.

Figure 11.1. If part of a vessel cannot be seen, assume it is dirty.

If the previous contents were poisonous, we should assume there is some poisonous material out of sight and breathing apparatus should be worn for entry.

Gas tests alone are not conclusive. There may be some sludge present which gives off gas when heated or disturbed.

(b) The last remark is illustrated by the following incident. To clean a paint mixing tank it was washed out with xylene. This cleaned the sides but some residue had to be scraped from the bottom. While an employee was doing so, wearing neither breathing apparatus nor a life-line, he was overcome by xylene which was trapped in the residue and escaped when it was disturbed (3).

(c) After a permit had been issued to weld inside a vessel a foreman noticed a deposit on the walls. He scraped some off, tested it and found that it burned. The permit was withdrawn.

(d) A tank had to be entered for inspection. It had contained only water and was not connected to any other equipment so the usual tests were not carried out. Three men went into the tank and were overcome. Two recovered but one died. The atmosphere inside the tank was tested afterwards and found to be deficient in oxygen. It is probable that rust formation used up some of the oxygen.

Section 5.3.4 describes how a similar effect caused a tank to collapse.

Never take short cuts in entering a vessel. Follow the rules.

See also Section 11.6 (b).

(e) Flammable or toxic liquids have been trapped inside the bearings of stirrers and have then leaked out. In one case a worker was overcome while working on a bearing although the vessel had been open for entry for 17 days. He disturbed some trapped liquid. Before issuing an entry permit look for any places in which liquid might be trapped. Vessels should always be slip-plated as close to the vessel as possible and on the vessel side of isolation valves. Otherwise liquid may be trapped between the valve and the slip-plate.

11.2 HAZARDOUS MATERIALS INTRODUCED

Sometimes, after a vessel has been freed from hazardous materials, they are then deliberately reintroduced, as in the following incidents.

(a) Two men went into a reactor to carry out a dye-penetrant test on a new weld using trichlorethylene. Because the weld was 8 m long the solvent was soon used up, and the man who was on duty at the entrance was asked to go for some more. He was away for 10 minutes. When he returned the two men inside the reactor had collapsed. Fortunately they were rescued and soon recovered.

The amount of solvent that can be taken into a vessel for dye-penetrant testing or other purposes should be limited so that evaporation of the complete amount will not bring the concentration above the TLV, making allowance for the air flow if the vessel is force-ventilated.

Stand-by workers should not leave a vessel when others are inside it.

(b) A most incredible case has been reported by OSHA (4). It was decided to shrink fit a bearing onto a shaft. The shaft was cooled—in a pit—by hosing liquefied petroleum gas onto it while the bearing was heated with an acetylene torch on the floor above the pit. An explosion occurred killing one man and injuring two others.

(c) The same OSHA report also describes several fatal fires and explosions which occurred while the insides of vessels were being painted, sometimes by spraying. In many cases the "cause" was said to be unsuitable lighting. But people should never be asked to work in a flammable atmosphere in view of the ease with which sources of ignition can turn up. The concentration of flammable vapor should never exceed 20% of the lower flammable limit while workers are in a vessel. And if necessary the atmosphere should be monitored continuously. Other fires or explosions were the result of leaks from welding equipment, often ignited when welding started again. Gas tests should always be carried out before welding starts. If oxygen is being used, then the atmosphere should be tested for oxygen as well as flammable vapors.

11.3 VESSELS NOT ISOLATED FROM SOURCES OF DANGER

Before entry is allowed into a vessel or other confined space the vessel should be isolated from sources of hazardous material by slip-plating or physically disconnecting all pipelines and by isolating all supplies of electricity, preferably by disconnecting the cables. On the whole these precautions seem to be followed. Accidents as the result of a failure to isolate are less common than those resulting from a failure to remove hazardous materials or from their deliberate reintroduction as described in Sections 11.1 and 11.2. However, the following are typical of the accidents that have occurred.

(a) A reactor had been isolated for overhaul. When maintenance was complete the slip-plates were removed and the vessel prepared for startup. It was then realized that an additional job had to be done. So workers were allowed to enter the vessel without the slip-plates being put back and without any gas tests being carried out. An explosion occurred, killing two and injuring two others. It was later found that hydrogen had leaked into the vessel through a leaking tube (5).

(b) The OSHA report describes a number of incidents in which steam lines failed, as the result of corrosion, while employees were working in a pit or other confined space from which they could not escape quickly. In general, steam lines, heating coils, etc., should be depressured and isolated before entry is permitted to a confined space (6).

(c) On a number of occasions people have been injured because machinery was started up while they were inside a vessel. For example, two workers were fixing new blades to No.2 unit in a pipe coating plant. A third worker wanted to start up another unit. By mistake he pressed the wrong button. No.2 unit moved and one of the workers was killed (7).

11.4 UNAUTHORIZED ENTRY

(a) Contractors, unfamiliar with a company's rules, have often entered vessels without authority. For example, a contractor's foreman was found inside a tank which was disconnected and open, ready for entry, but not yet tested. He had been asked to estimate the cost of cleaning the tank. The foreman said that he did not realize that a permit was needed just for inspection. He had been given a copy of the plant rules but had not read them.

   If vessels are open, but entry is not yet authorized, the manhole should be covered by a barrier. Do not rely on contractors reading rules. Explain the rules to them.

(b) It is not only contractors who enter vessels without authorization, as shown by Section 11.1 (d) and the following incident.

   A process foreman had a last look in a vessel before it was boxed up. He saw an old gasket lying on the floor. He decided to go in and remove it. Everyone else was at lunch, so he decided to go in alone. On the way out, while climbing a ladder, he slipped and fell and was knocked out. His tongue blocked his throat and he suffocated.

(c) The incident described in Section 12.3.2 (d) shows that you do not have to get inside a confined space to be overcome. Your head is enough. People should never put their heads inside a vessel unless entry has been authorized.

11.5 ENTRY INTO VESSELS WITH IRRESPIRABLE ATMOSPHERES

A man was standing on a ladder ready to go down into a drain manhole to plug one of the inlet lines. The drain contained some hydrogen sulphide so he had breathing apparatus ready. But he had not yet put on the facepiece because he was well outside the manhole. His feet were at ground level (Figure 11.2). He was about to put on a safety harness when his two companions heard a shout and saw him sliding into the manhole. They were unable to catch him and his body was recovered from the outfall. He had been overcome by hydrogen sulphide arising from the drains although his face was 1.5 m above ground level.

Figure 11.2. A man was overcome by fumes from the manhole.

This incident shows that if a vessel or confined space contains a toxic gas, people can be overcome a meter or more from the opening. Similar incidents involving nitrogen are described in Section 12.3.2 (b).

The incident and the one described in Section 12.3.2 (c) show that special precautions are necessary when entering vessels containing atmospheres that contain so much toxic gas or so little oxygen that there is immediate danger to life. In most cases of entry in which breathing apparatus is worn, it is used because the atmosphere is unpleasant to breath or will cause harm if breathed for several hours. Only on very rare occasions should it be necessary to enter vessels containing an irrespirable atmosphere. In such cases two persons trained in rescue and resuscitation should be on duty outside the vessel. And they should have available any equipment necessary for rescuing the person inside the vessel, who should always be kept in view.

11.6 RESCUE

If we see another person overcome inside a vessel, there is a very strong natural impulse to rush in and rescue him, even though no breathing apparatus is available. Misguided bravery of this sort can mean that other people have to rescue two people instead of one, as shown by the following incidents and by Section 12.3.3.

(a) A contractor entered the combustion chamber of an inert gas plant watched by two standby men but without waiting for the breathing apparatus to arrive. While he was climbing out of the chamber he lost consciousness halfway up. His body was caught between the ladder and the chamber wall. The standby men could not pull him out with the lifeline to which he was attached. So one of the standby men climbed in to try to free him, without breathing apparatus or a lifeline. The standby man also lost consciousness. The contractor was finally pulled free and recovered. The standby man was rescued by the fire service but by this time he was dead.

(b) Three men were required to inspect the ballast tanks on a barge tied up at an isolated wharf 20 km from the plant. No tests were carried out. One tank was inspected without incident. But on entering the second tank, the first man collapsed at the foot of the ladder. The second man entered to rescue him and also collapsed. The third man called for assistance. Helpers who were asked to assist in recovering the two men were partly overcome themselves. Representatives of the safety department 20 km away set out with breathing apparatus. One man died before he could be rescued.

Tests on other tanks showed oxygen contents as low as 5%. It is believed that rust formation had used up the oxygen, as in the incident described in Section 11.1 (d) (8).

11.7 ANALYSIS OF VESSEL ATMOSPHERE

The following incidents show how errors in analysis nearly resulted in people entering an unsafe atmosphere. In both cases the laboratory staff were asked to test the atmosphere inside a vessel. In the first case they checked the oxygen content with a portable analyzer in which a sample of the atmosphere being tested is drawn through the apparatus with an aspirator. There was a blockage in the apparatus, so it merely registered the oxygen content of the air already inside it. A bubbler or other means of indicating flow should have been fitted.

In the second case the sample was taken near the manhole instead of being taken near the middle of the vessel. Samples should always be taken well inside the vessel. Long sample tubes should be available so that this can be done. In large vessels and in long, tortuous places like flue gas ducts, samples should be taken at several points.

On both these occasions vigilance by the operating staff prevented what might have been a serious incident. They suspected that something was wrong with the analysis results and investigated the way in which the samples were taken.

Inevitably, a book like this one is a record of failures. It is pleasant to be able to describe accidents prevented by the alertness of operating staff.

REFERENCES

1. T. A. Kletz in H. H. Fawcett and W. S. Wood (editors), *Safety in Chemical Operations,* 2nd edition, 1982, Chapter 36.
2. F. P. Lees, *Loss Prevention in the Process Industries,* Butterworths, 1980, Chapter 21.
3. *Health and Safety at Work,* July 1982, p. 38.
4. W. W. Cloe, *Selected Occupational Fatalities related to Fire and/or Explosion in Confined Work Spaces as found in Reports of OSHA Fatality/Catastophe Investigations,* Report No. OSHA/RP-82/002, U.S. Dept. of Labor, Washington DC, April 1982, p. 32.
5. Reference 4, p. 22.
6. Reference 4, pp. 65–71.
7. *Occupational Safety and Health,* Supplement, Oct. 1975.
8. *Petroleum Review,* May 1977, p. 49.

Chapter 12

Hazards of Common Materials

This chapter is not concerned with the hazards of obviously dangerous materials such as highly flammable liquids and gases, or toxic materials. Rather, the focus is on accidents involving those common but dangerous substances: air, water, nitrogen and heavy oils.

12.1 COMPRESSED AIR

Many operators find it hard to grasp the power of compressed air. Section 2.2 (a) described how the end was blown off a pressure vessel, killing two men, because the vent was choked. Compressed air was being blown into the vessel, to prove that the inlet line was clear. It was estimated that the gauge pressure reached 20 psi (1.3 bar) when the burst occurred. The operators found it hard to believe that a pressure of "only 20 pounds" could do so much damage. Explosion experts had to be brought in to convince them that a chemical explosion had not occurred.

Unfortunately, operators often confuse a force (such as 20 pounds) with a pressure (such as 20 pounds per square inch) and forget to multiply the 20 pounds by the number of square inches in the end of the vessel.

A similar accident is described in Section 13.5. Other incidents in which equipment was damaged by compressed air are described in Section 5.2.2. Because employees do not always appreciate the power of compressed air, it has sometimes been used to remove dust from workbenches or clothing. And dust has been blown into people's eyes or into cuts in the skin. Worse still, compressed air has been used for "horseplay." A man was killed when a compressed air hose was pushed up his rectum (1).

Fires have often occurred when air is compressed. Above 140°C lubricating oil oxidizes and forms a carbonaceous deposit on the walls of air compressor delivery lines. If the deposit is thin it is kept cool by conduction through the pipework. But when deposits get too thick they can catch fire. Sometimes the delivery pipe has gotten so hot that it has burst or the aftercooler has been damaged. In one case the fire vaporized some of the water in the aftercooler and set up a shock wave which caused serious damage to the cooling water lines.

To prevent fires or explosions in air compressors:

1. The delivery temperature should be kept below 140°C. It is easier to do this if the inlet filters are kept clean and the suction line is not throttled.
2. The delivery pipework should be kept clean by avoiding traps in which oil can collect and by regular cleaning. Deposits should not be allowed to get more than 1/8 inch thick (3 mm). One fire occurred in a compressor on which the pulsation dampers were so designed that it was impossible to clean them.
3. The use of special lubricants may reduce the formation of deposits. On some rotary air compressors a large oil surface is exposed to the air. Deposits can form and catch fire at temperatures below 140°C.

    Normally, after the aftercooler, the air is too cool for deposits to form or catch fire. But this may not always be the case. In one case an instrument air drier became contaminated with oil and caught fire during the drying cycle.

    Other hazards of compressed air are described in Reference 2.

12.2 WATER

The hazards of water hammer are described in Section 9.1.5 and the hazards of ice formation in Section 9.1.1. This section describes some accidents that have occurred as the result of the sudden vaporization of water, incidents known as slopovers, boilovers, foamovers or puking. Sections 9.1.1 and 12.4.5 describe incidents in which vessels burst because water which had collected in a trap was suddenly vaporized. But most slopovers have occurred when a water layer in a tank was suddenly vaporized, as in the following incident:

(a) Hot oil, the residue from a batch distillation, was being moved into a heavy residue storage tank. There was a layer of water in the tank—the result of steaming the oil transfer line after previous movements—and this vaporized with explosive violence. The roof of the tank was lifted and structures over 20 m tall were covered with black oil. A man who saw the incident said that the tank exploded and it was an explosion—a sudden release of energy—though due to physical rather than chemical causes.

To prevent similar incidents from happening, if heavy oil is being transferred into a tank, incoming oil should be kept *below* 100°C and a high temperature alarm should be installed on the oil line. Alternatively, water should be drained from the tank, the tank kept *above* 100°C, and the tank contents circulated before the movement of oil into the tank starts. In addition, the movement of oil into the tank should start at a low rate.

(b) In other cases a water layer has vaporized suddenly when it was heated by conduction from a hotter oil layer above. For example, to clean a heavy oil tank some lighter oil was put into it and heated by the steam coil. There was a layer of water below the oil. The operators were told to keep the temperature of the oil below 100°C. But they did not realize that the height of the thermocouple (1.5 m) was above that of the top of the oil (1.2 m). Although the thermocouple was reading 77°C the oil was above 100°C, the water vaporized and the roof was blown off the tank.

As the water started to boil and lift up the oil, the hydrostatic pressure on the water was reduced and this caused the water to boil with greater vigor.

(c) Some paraffin which had been used for cleaning was left in a bucket. There was some water under the paraffin. Some hot equipment set fire to some cleaning rags and the fire spread to the paraffin in the bucket.

To put out the fire, a man threw a shovelfull of wet scale into the bucket. The water became mixed with the oil, turned to steam and blew the oil over the man who was standing 1–2 m away. He died from his burns.

1. Never mix water and hot oil.
2. Do not use flammable solvents for cleaning.
3. Do not carry flammable liquids in buckets. Use a closed can. See Section 7.1.3.

Other hazards of water are described in Reference (3).

12.3 NITROGEN (4)

Nitrogen is widely used to prevent the formation of flammable mixtures of gas or vapor and air. Flammable gases or vapors are removed with nitrogen before air is admitted to a plant, and air is removed with nitrogen before flammable gases or vapors are admitted.

There is no doubt that without nitrogen (or other inert gas), many more people would be killed by fire or explosion. Nevertheless we have paid a heavy price for the benefits of nitrogen. Many people have been asphyxiated by it. In one group of companies in the period 1960–1978, 13 employees were killed by fire or explosion, 13 by toxic or corrosive chemicals and 7 by nitrogen. It is our most dangerous gas.

This section describes some accidents in which people were killed or overcome by nitrogen. Some of the accidents occurred because nitrogen was used instead of air. In others people were unaware of the dangers of nitrogen or were not aware that it was present.

The name "inert gas," often used to describe nitrogen, is misleading. It suggests a harmless gas. Nitrogen is not harmless. If a person enters an atmosphere of nitrogen he can lose consciousness, without any warning symptoms or distress, in as little as 20 seconds. Death can follow in three or four minutes. A person falls as if struck down by a blow on the head.

12.3.1 NITROGEN CONFUSED WITH AIR

Many accidents have occurred because nitrogen was used instead of compressed air. For example, on one occasion a control room operator noticed a peculiar smell. On investigation it was found that a hose, connected to a nitrogen line, had been attached to the ventilation intake. This had been done to improve the ventilation of the control room which was rather hot. On other occasions nitrogen has been used by mistake to freshen the atmosphere in vessels in which employees were working.

On another occasion nitrogen was used by mistake to power an air-driven light, used during entry to a vessel. In this case the error was discovered in time.

More serious are incidents in which nitrogen has been connected to breathing apparatus.

To prevent these errors many companies use different fittings for compressed air and nitrogen. Nevertheless, confusion can still occur as the following story shows:

An operator donned a fresh air hood to avoid breathing harmful fumes. Almost at once he felt ill and fell down. Instinctively he pulled off the hood and quickly recovered. It was then found that the hood had been connected by mistake to a supply of nitrogen instead of compressed air.

Different connections were used for nitrogen and compressed air, so it was difficult at first to see how a mistake had been made.

However, the place where the man was working was a long way from the nearest compressed air connection so several lengths of hose had to be joined together. This was done by cutting off the special couplings and using simple nipples and clamps. Finally, the hoses were joined to one projecting through an opening in the wall of a warehouse. The operator then went into the warehouse, selected what he thought was the other end of the projecting hose and connected it to the airline. Unfortunately, there were several hoses on the floor of the warehouse and the one to which he had joined the airline outside was already connected to a nitrogen line.

To prevent incidents similar to those described, we should:

1. Use cylinder air for breathing apparatus
2. Label all service points
3. Use different connections for air and nitrogen and publicize the difference so that everyone knows.

12.3.2 IGNORANCE OF THE DANGERS

(a) A member of a cleaning crew decided to recover a rope which was half inside a vessel and was caught up on something inside. While kneeling down, trying to disentangle the rope, he was overcome by nitrogen. Afterward he admitted that if necessary he would have entered the vessel.

(b) On several occasions people who were working on or near leaky joints on nitrogen lines have been affected. Although they knew that nitrogen was harmful, they did not consider that the amount coming out of a leaky joint would harm them.

(c) Two men without masks were killed because they entered a vessel containing nitrogen. Possibly they had removed their masks on other occasions, when the atmosphere was not harmful to breathe for a moment of two, and did not appreciate that in a 100% nitrogen atmosphere they would be overcome in seconds. It is believed that one man entered the vessel, removed his mask and was overcome and that the second man then entered, without a mask, to rescue him.

   Entry should not normally be allowed to vessels containing irrespirable atmospheres. Special precautions are necessary if entry is permitted. See Section 11.5.

(d) You do not have to get right inside a confined space to be overcome. Your head is enough.

   When a plant was being leak tested with nitrogen after a shutdown, a leak was found on a manhole joint on the side of a vessel.

The pressure was blown off and a fitter asked to remake the joint. While he was doing so the joint ring fell into the vessel. Without thinking, the fitter squeezed the upper part of his body through the manhole so that he could reach down and pick up the joint.

His companion saw his movements cease, realized he was unconscious and pulled him out into the open air where he soon recovered.

(e) In another incident the cover of a large converter was removed but nitrogen was kept flowing through it to protect the catalyst. An inspector did not ask for an entry permit, as he intended only to "peep in." Fortunately someone noticed that he had not moved for a while and he was rescued in time.

12.3.3 NITROGEN NOT KNOWN TO BE PRESENT

Some of the incidents described in Section 12.3.2 may fall into this category. Most of the incidents of this type, however, have occurred during construction when one group of workers have connected up the nitrogen supply to a vessel unknown to others. The following is an account of a particularly tragic accident of this type.

Instrument personnel were working inside a series of new tanks, installing and adjusting the instruments. About eight weeks earlier a nitrogen manifold to the tanks had been installed and pressure tested; the pressure was then blown-off and the nitrogen isolated by a valve at the plant boundary.

The day before the accident the nitrogen line was put back up to pressure because the nitrogen was required on some of the other tanks.

On the day of the accident an instrument mechanic entered a 2 m³ tank to adjust the instruments. There was no written entry permit because the people concerned believed, mistakenly, that they were not required in a new plant until water or process fluids had been introduced. Although the tank was only six feet tall and had an open manhole at the top, the mechanic collapsed. An engineer arrived at the vessel about five minutes later to see how the job was getting on. He saw the first man lying on the bottom, climbed in to rescue him and was overcome as soon as he bent down.

Another engineer arrived after another five or ten minutes. He fetched the process supervisor and then entered the vessel. He also collapsed. The supervisor called the plant fire service. Before they arrived the third man recovered sufficiently to be able to climb out of the vessel. The second man was rescued and recovered, but the first man died.

It is believed that an hour or two before the incident somebody opened the nitrogen valve leading to the vessel and then closed it.

Whan can we learn from this incident?

1. If someone is overcome inside a vessel or pit we should never attempt to rescue him without breathing apparatus. We must curb our natural human tendency to rush to his aid or there will be two people to rescue instead of one. See Section 11.6.
2. Once a vessel has been connected up to any process or service line, the full permit-to-work and entry procedure should be followed. In the present case, this should have started eight weeks before the incident. And the nitrogen line should have been disconnected or slip-plated where it entered the vessel.

   There should be a formal handover from construction so that everyone is aware when it has taken place. The final connection to process or service lines is best made by plant fitters rather than by the construction team. In each plant, the procedure for handover should be described in a plant instruction.
3. When the plant is still in the hands of construction the normal permit-to-work procedure is not necessary but an entry permit system should be in force. Before anyone enters a vessel, it should be inspected by a competent, experienced person who will certify that it is isolated and free from danger. When a tank is being built, at a certain height of walls (say, equal to the diameter) it should be deemed to be a confined space and the entry procedure should apply.
4. All managers and supervisors should be aware of the procedure for handover and entry to vessels.

12.4 HEAVY OILS (INCLUDING HEAT TRANSFER OILS)

This term is used to describe oils which have a flash point above ambient temperature. They will therefore not burn or explode at ambient temperature but will do so when hot. Unfortunately many people do not realize this and treat heavy oils with a disrespect that they would never apply to gasoline, as shown by the incidents described below. Another incident was described in Section 12.2 (c). Heavy oils are widely used as fuel oils, solvents and heat transfer oils as well as process materials.

12.4.1 TRACES OF HEAVY OIL IN EMPTY TANKS

Repairs had to be carried out to the roof of a storage tank which had contained heavy oil. The tank was cleaned out as far as possible and two welders started work. They saw smoke coming out of the vent and flames coming out of the hole they had cut. They started to leave, but before they could do so the tank's roof lifted and a flame 25 m long came out. One of the men was killed and the other was badly burned. The residue in the tank continued to burn for 10–15 minutes (5).

Though the tank had been cleaned, traces of heavy oil were stuck to the sides or behind rust or trapped between plates. These traces of oil were vaporized by the welding and ignited.

Some old tanks are welded along the outside edge of the lap only, thus making a trap from which it is hard to remove liquids. Even light oils can be trapped in this way. See Section 5.4.2 (c) and Figure 5.9.

A similar incident is described in an official report (6). A tank with a gummy deposit on the walls and roof had to be demolished. The deposit was unaffected by steaming but gave off vapor when a burner's torch was applied to the outside. The vapor exploded, killing six firemen who were on the roof at the time.

It is almost impossible to completely clean a tank (or other equipment) which has contained heavy oils, residues or polymers or material that is solid at ambient temperature, particularly if the tank is corroded.

Tanks which have contained heavy oils are more dangerous than tanks which have contained lighter oils such as gasoline. Gasoline can be completely removed.

Note also that while light oils such as gasoline can be detected with a combustible gas detector, heavy oils cannot be detected. Even if a heavy oil is heated above its flash point the vapor will cool down in the detector before it reaches the sensitive element.

Before welding is allowed on tanks which have contained heavy oils, the tanks should be filled with inert gas or with fire-fighting foam generated with inert gas, *not* with fire-fighting foam generated with air. Filling the tank with water can reduce the volume to be inerted.

12.4.2 TRACES OF HEAVY OIL IN PIPELINES

Some old pipelines had to be demolished. They were cleaned as far as possible and then tested with a combustible gas detector. No gas or vapor was detected so a burner was given permission to cut them up. While doing so, sitting on the pipes 4 m above the ground, a tarry substance seeped out of one of the pipes and caught fire. The fire spread to the burner's clothing and he ended up in the hospital with burns to his face and legs.

The deposit did not give off enough vapor when cold for it to have been detected by the combustible gas detector.

It is almost impossible to completely clean pipes which have contained heavy oils or polymers. When demolishing old pipelines there should be as many open ends as possible so that pressure cannot build up. And good access should be provided so that the burner or welder can escape readily if he needs to do so.

12.4.3 POOLS OF HEAVY OIL

An ore-extracting process was carried out in a building with wooden floors. But this was considered safe because the solvent used had a flash point of 42°C and it was used cold. Leaks of solvent drained into a pit inside the building. While welding was taking place, a burning piece of rag fell into the pit and in a few seconds the solvent film which covered the water in the pit was on fire. The rag acted as a wick and set fire to the solvent, although a spark or a match would not have done so. The fire spread to the wooden floor, some glass pipes burst and these added more fuel to the fire. In a few minutes the building was ablaze and two-thirds of the contents were destroyed (7).

12.4.4 SPILLAGES OF HEAVY OIL, INCLUDING SPILLAGES ON INSULATION

The heat transfer section of a plant was filled with oil after maintenance by opening a vent at the highest point and pumping oil into the system until it overflowed out of the vent. The overflow should have been collected in a bucket but sometimes a bucket was not used or the bucket was overfilled. Nobody worried about small spillages because the flash point of the oil was above ambient temperature and its boiling point and auto-ignition temperature were both above 300°C.

A month after such a spillage the oil caught fire. Some of it might have soaked into insulation and, if so, this would have caused it to degrade, lowering its auto-ignition temperature so that it ignited at the temperature of the hot pipework.

The oil fire caused a leak of process gas which exploded causing further localized damage and an oil fire.

All spillages, particularly those of high boiling liquids, should be cleaned up promptly. Light oils will evaporate but heavy oils will not. Besides the fire hazard, spillages produce a risk of slipping.

Insulation which has been impregnated with heavy oil—or any other organic liquid—should be removed as soon as possible before it ignites. If oil is left in contact with insulation materials, the auto-ignition temperature is lowered by 100–200°C (8).

12.4.5 HEAVY OIL FIREBALLS

Sections 9.1.2 and 12.2 described incidents which occurred when heavy oils, at temperatures above 100°C, came into contact with water. The water vaporized with explosive violence and a mixture of steam and oil was blown out of the vessel, after rupturing it.

In another incident of the same nature the oil caught fire.

A furnace supplied heat transfer oil to four reboilers. One was isolated for repair and then pressure tested. The water was drained out of the shell but the drain valve was eight inches above the bottom tube plate and so a layer of water was left in the reboiler (Figure 12.1).

When the reboiler was brought back on line the water was swept into the heat transfer oil lines and immediately vaporized. This set up a liquid hammer which burst the surge tank. It was estimated that this required a gauge pressure of 450 psi (30 bar). The top of the vessel was blown off in one piece and the rest of the vessel was split into 20 pieces. The hot oil formed a cloud of fine mist which ignited immediately forming a fireball 35 m in diameter.

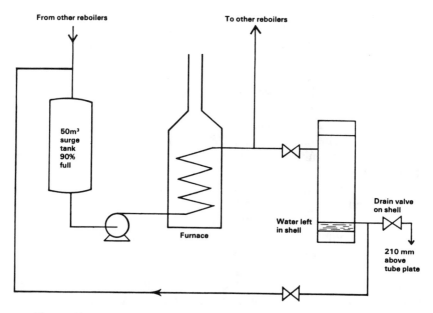

Figure 12.1. Water left in the heat exchanger was vaporized by hot oil.

Recommendations which followed from this incident are:

1. Adequate facilities must be provided for draining water from heat transfer and other hot oil systems.
2. Oil rather than water should be used for pressure testing.
3. Surge vessels should operate about half full, not 90% full as in this case.
4. In new plants, water should be considered as a heat transfer medium instead of oil. A decision to use water has to be made early in the design because the operating pressure will be higher. Although this will add to the cost, there will be a saving in lower fire protection costs. In some plants the heat transfer oil is a bigger fire hazard than the process materials (9–11).

REFERENCES

1. *Safety Management* (S. Africa), April 1982, p. 30.
2. *Hazard of Air*, American Oil Company, 5th edition, 1964.
3. *Hazard of Water*, American Oil Company, 5th edition, 1964.
4. T. A. Kletz, *Nitrogen—our most dangerous gas, Proceedings of the 3rd International Symposium on Loss Prevention and Safety Promotion in the Process Industries*, Swiss Society of Chemical Industries, 1980, p. 1518.
5. T. A. Kletz, *J of Hazardous Materials*, Vol. 1, No. 2, 1976, p. 165.
6. A. W. M. Davies, *Public Enquiry into a Fire at Dudgeon's Wharf on 17 July 1969*, London, Her Majesty's Stationery Office, 1970.
7. R. Hoy-Petersen, *Proceedings of the First International Symposium on Loss Prevention and Safety Promotion in the Process Industries*, Elsevier, Amsterdam, 1984, p. 325.
8. P. E. Macdermott, *Petroleum Review*, July 1976.
9. *A Guide to Effective Industrial Safety*, Jack W. Boley, 1977; Gulf Publishing Co., Houston.
10. *Inspection of Chemical Plants*, Leslie Pilborough, 1977; Gulf Publishing Co., Houston.
11. *Unconfined Vapor Cloud Explosions*, Keith Gugan, 1979; Gulf Publishing Co., Houston.

Chapter 13

Tank Trucks and Cars

This chapter is not concerned with accidents on the road. Rather, it describes some of the many incidents that have occurred while tank trucks and cars (known in Europe as road and rail tank wagons) were being filled or emptied.

13.1 OVERFILLING

Tank trucks and cars have been overfilled on many occasions both when filled automatically and when filled by hand.

In automatic systems the filler sets the quantity to be filled on a meter which closes a valve when this required quantity has been delivered. Overfilling has occurred because the wrong quantity was set on the meter, because there was already some liquid in the tank (left over from the previous load) or because the filling equipment failed. For these reasons many companies now fit their tank trucks with high-level trips which automatically close a valve in the filling line.

Tank trucks and cars which are filled by hand have been overfilled because the filler left the job for a few minutes and returned too late.

On one occasion an operator thought that a tank truck had a single-compartment tank when in fact there were two compartments. He tried to put the full load into one compartment.

On another occasion, after a tank truck had been filled during the night, the operator completed a "filling certificate"—a very small piece of paper—and slipped it inside the dispatch papers. This was the usual practice.

When the next shift came on duty the driver had not returned to get the truck. The overnight record sheets had all been sent to the plant office.

So the operator shook the dispatch papers to see if there was a filling certificate among them. Nothing fell out because the certificate was caught up in the other papers.

The operator therefore started to fill the tanker again.

In contrast, a case of overfilling, which was the subject of an official report (1), was due to the poor design of complex automatic equipment at a large terminal for loading gasoline and other hydrocarbons.

The grade and quantity of product required were set on a meter. The driver inserted an authorization card and pressed the start button. The required quantity was then delivered automatically. The filling arm had to be lowered before filling could start.

One day the automatic equipment broke down and the foreman decided to change over to manual filling. He asked the drivers to check that the hand valves on the filling lines were shut but he did not check himself. He then operated the switches that opened the automatic valves. *Some of the hand valves were open.* Gasoline and other products came out, overfilled the tanker (or splashed directly on the ground) and caught fire. Three men were killed, eleven injured and the whole row of 18 filling points was destroyed.

To quote from the official report, "The decision to override the individual controls on the loading arms by means of a central switchboard, without the most rigid safeguards, was a tragic one. After its installation an accident from that moment on became inevitable sooner or later.

"That this switchboard was installed, with the approval of the terminal management . . . in a switchroom from which the loading stands were not visible, suggests some failure to take into account the basic fundamentals of safety in operation of plant.

". . . had the same imagination and the same zeal been displayed in matters of safety as was applied to sophistication of equipment and efficient utilization of plant and men, the accident need not have occurred."

13.2 BURST HOSES

Hoses have failed while tank trucks or cars were being filled or emptied for all the reasons listed in Section 7.1.6, in particular because damaged hoses or hoses made from the wrong material were used. However, the commonest cause of hose failure is the tanker driving away before the hose is disconnected.

The following incidents are typical:

(a) A tank truck was left at a plant for filling with liquefied flammable gas. Some hours later the transport foreman assumed that it would be ready and sent a driver to get it. There was no one in the plant office so the driver went to the loading bay. He found that the truck was grounded and that the grounding lead had been looped through the steering wheel—the usual practice—to prevent the driver from driving away before disconnecting it. He removed the lead and drove off, snapping off the filling branch and tearing the hose that was connected to the vent line. Fortunately there was no flow through the filling line at the time though the valves were open, and the spillage was relatively small. It did not catch fire.

Plant instructions stated that a portable barrier should be put in front of tank trucks which were filled but it was not being used. However, if it had been in use, the driver might have removed it.

A device that can be fitted to a tank truck to prevent anyone driving it away while a hose is connected is described in Reference 2. A plate is fixed in front of the hose connection. To connect the hose this plate has to be moved aside and this applies the brakes. Reference 3 describes a special type of hose that seals automatically if it breaks.

Remotely operated emergency isolation valves (see Section 7.2.1) should be fitted on filling lines. If the hose breaks for any reason the flow can be stopped by pressing a button located at a safe distance. Back flow from the tank truck or car can be prevented by a nonreturn valve.

Note that it is not necessary to ground tank trucks containing liquefied flammable gases because no air is present in the tank.

(b) Gasoline was being discharged at a service station from a tank truck which was carrying diesel fuel in one compartment. To save time the driver decided to discharge the diesel fuel while discharging the gasoline. To do this he had to move the tank truck about 1 or 2 meters.

He drove the truck slowly forward, while the discharge of fuel continued. The hose caught on an obstruction and was pulled part way out of its fastening. Gasoline escaped and caught fire. The service station and tank truck were destroyed (4).

13.3 FIRES AND EXPLOSIONS

A number of explosions or fires have occurred in tank trucks or cars while they were being filled. The most common cause is "switch filling." A tank contains some flammable vapor, such as gasoline vapor, from a previous load and is then filled with a safer, higher boiling liquid such as gas oil. The gas oil is not flammable at ambient temperature. So no special precautions are normally necessary to prevent the formation of static electricity. The tank may be filled quickly, may even be splash-filled, a static charge is formed and a spark jumps from the liquid to the wall of the tank, igniting the gasoline vapor.

A similar incident occurred in a tank truck used to carry waste liquids. While it was being filled with a nonflammable liquid and the driver was standing on the top, smoking, an explosion occurred and the manhole cover was thrown 60 m.

On its previous journey the tank truck had carried a waste liquid containing dissolved flammable gas. Some of the gas was left in the tank and was pushed out when it was filled with the next load.

Flammable liquids should never be splash-filled, even though they are below their flash points. The splash filling may form spray which can be ignited by a static discharge. Sprays, like dusts, can be ignited at any temperature.

On one occasion a tank truck was being splash-filled with gas oil, flash point 60°C. The splashing produced a lot of mist and it also produced a charge of static electricity on the gas oil. This discharged, igniting the mist. There was a fire with flames 10 m high, but no explosion. The flames went out as soon as the mist had been burned.

Many thousands of tank trucks had been splash-filled with gas oil at this installation before conditions were exactly right for a fire to occur. When handling flammable gases or liquids, we should never say, "It's O.K., we've been doing it this way for 20 years and never had a fire." Such a statement should be made only if an explosion in the 21st year is acceptable.

Note that grounding a tank truck will not prevent ignition of vapor by a discharge of static electricity. Grounding will prevent a discharge from the tank to earth but it will not prevent a discharge from the liquid in the tank to the tank or to the filling arm.

For more information on static electricity see Chapter 15.

13.4 LIQUEFIED FLAMMABLE GASES

Tank trucks or cars which carry liquefied gases under pressure at ambient temperature present additional hazards.

When the tanks are filled, the vapor is vented to a stack or back to the plant through a vapor return line which is fitted to the top of the tank. An official report (5) described a fire that occurred because the fillers had not bothered to connect up this vapor return line. Vapors were discharged into the working area. Seven people were injured.

Following this incident, a survey at another large installation showed that the fillers there were also forgetting to connect up the vapor lines. Reference 5 also reports that at another plant the vapor return line was connected in error to another filling line. The vapor could not escape, the pressure in the tank rose and the filling hose burst. There was no emergency isolation valve in the filling line, no nonreturn valve on the tank [see Section 13.2 (a)] and no excess flow valve on either, so the spillage was substantial.

Vapor return lines and filling lines should be fitted with different sizes or types of connections.

Section 13.2 (a) described another incident with LFG.

13.5 COMPRESSED AIR

Compressed air is often used to empty tank trucks and cars. Plastic pellets are one of the loads that are often blown out of tank trucks. When the tank is empty the driver vents the tank and then looks through the manhole to check that the tank is empty. One day a driver who was not regularly employed on this job started to open the manhole before releasing the pressure. When he had opened two out of five quick-release fastenings, the manhole blew open. The driver was blown off the tank top and killed.

Either the driver forgot to vent the tank or thought it would be safe to let the pressure (a gauge pressure of 10 psi or 0.7 bar) blow off through the manhole. After the accident the manhole covers were replaced by a different type in which two movements are needed to release the fastenings. The first movement allows the cover to be raised about 1/4 inch while still capable of carrying the full pressure. If the pressure has not been blown off this is immediately apparent and the cover can be resealed or the pressure allowed to blow off.

In addition the vent was repositioned at the foot of the ladder (6).

Many of those concerned were surprised that a pressure of "only 10 pounds" could cause a man to be blown off the top of the tank. They forgot that 10 psi is not a small pressure. It is 10 pounds force on every square inch. See Section 12.1.

A similar incident is described in Section 17.1.

13.6 TIPPING UP

On several occasions tank trailers have tipped up because the rear compartments were emptied first, as shown in Figure 13.1.

It is not always possible to keep the trailer connected to the truck's unit during loading/unloading. If it is not connected the front compartments should be filled last and emptied first or a support put under the front of the trailer.

Figure 13.1. A tank trailer may tip up if the rear compartments are emptied first.

13.7 EMPTYING INTO OR FILLING FROM THE WRONG PLACE

On many occasions tank trucks have been discharged into the wrong tank. The following incident is typical of many.

A tank truck containing isopropanol arrived at a plant during the night. It was directed to a unit which received regular supplies by tank trucks. They were expecting a load of ethylene glycol. So without looking at the label or the delivery note they emptied the tank truck into the ethylene glycol tank and contaminated 100 tons of ethylene glycol.

Fortunately in this case the two materials did not react. People who have emptied acid into alkali tanks have been less fortunate.

A plant received caustic soda in tank cars and acid in tank trucks. One day a load of caustic acid arrived in a tank truck. It was labelled "caustic soda" and the delivery papers said that it was caustic soda. But the operators were so used to receiving acid by rail that they spent two hours making an adaptor to enable them to pump the contents of the tank truck into the acid tank.

On other occasions tank trucks have been filled with the wrong material. In particular, liquid oxygen or liquid air has been supplied instead of liquid nitrogen. One incident, the result of confusion over labeling, was described in Section 4.1 (f). Liquid nitrogen should always be analyzed before it is off-loaded.

The author does not know of any case in which delivery of liquid oxygen instead of liquid nitrogen caused an explosion. But in one case the "nitrogen" was used to inert a catalyst bed and the catalyst got hot; in another case a high oxygen concentration alarm in the plant sounded and in several cases check analyses showed that oxygen had been supplied.

Many suppliers of liquefied gases state that they use different hose connections for liquid oxygen and liquid nitrogen so mistakes cannot arise. However, mistakes *have* occurred, possibly because of the well-known tendency of operators to acquire adaptors.

The following incident involved cylinders rather than bulk loads but it shows how alertness to an unusual observation can prevent an accident.

A plant used nitrogen in large cylinders. One day a cylinder of oxygen, intended for another plant, was delivered in error. The foreman noticed that the cylinder had an unusual color and unusual fittings and he thought it strange that only one was delivered. Usually several cylinders were delivered at a time. Nevertheless he accepted the cylinder. He did not notice that the invoice said "oxygen."

The invoice was sent as usual to the purchasing department for payment. The young clerk who dealt with it realized that oxygen had been delivered to a unit that had never received it before. She told her supervisore who telephoned the plant and the error came to light.

Another success story. See Section 11.7.

13.8 CONTACT WITH LIVE POWER LINES

The manhole covers on tank cars are sometimes sealed with wires. Loose ends of wires protruding above the manhole cover have come into contact with the overhead electric wire which supplies power to the train and caused a short circuit.

In the UK there is normally a gap of four inches between the highest point of the tank car and the lowest point of the cables, but if the gap falls below two inches arcing may occur (7).

Somewhat similar incidents have occurred on railway lines powered by a third electrified rail. The cap covering the discharge pipe has vibrated loose, the retaining chain has been too long and the cap has contacted the third rail (7).

REFERENCES

1. H. K. Black, *Report on a Fatal Accident and Fire at the West London Terminal on 1 April 1967,* London, Her Majesty's Stationery Office, London, 1967.
2. T. A. Kletz, *Loss Prevention,* Vol. 10, 1976, p. 151.
3. *Petroleum Review,* July 1976, p. 433.
4. *Petroleum Review,* July 1976, p. 428.
5. *Annual Report of H.M. Inspectors of Explosives for 1967,* Her Majesty's Stationery Office, London, 1968.
6. T. A. Kletz, *Loss Prevention,* Vol. 13, 1980, p. 1
7. *Petroleum Review,* April 1976, p. 241.

Chapter 14

Testing of Trips and Other Protective Systems

Many accidents have occurred because instrument readings or alarms were ignored. See Sections 3.2.8, 3.3.1 and 3.3.2. Many other accidents have occurred because alarms and trips were not tested, or not tested thoroughly, or because alarms and trips were made inoperative, or their settings altered, without authority. These and some related accidents are described below.

14.1 TESTING SHOULD BE THOROUGH

All protective equipment should be tested regularly or it may not work when required. While it is sufficient to test relief valves every year or every two years, instrumented alarms and trips are less reliable and should be inspected every month or so.

Testing must be thorough and as near as possible to real life situations, as shown by the following incidents:

(a) A high temperature trip on a furnace failed to operate. The furnace was seriously damaged. The trip did not operate because the pointer touched the plastic front of the instrument case and this prevented it moving to the trip level. The instrument had been tested regularly—by injecting a current from a potentiometer—but to do this *the instrument was removed from its case and taken to the workshop*.

(b) A reactor was fitted with a high temperature trip which closed a valve in the feed line. When a high temperature occurred, the trip valve failed to close although it had been tested regularly.

161

Investigation showed that the pressure drop through the trip valve—a globe valve—was so high that the valve could not close against it. There was a flow control valve in series with the trip valve (Figure 14.1) and the trip normally closed this valve as well. However, this valve failed in the open position—this was the reason for the high temperature in the reactor—and the full upstream pressure was applied to the trip valve.

Emergency valves should be tested against the maximum pressure or flow they may experience and, whenever possible, should be installed so that the flow assists closing.

14.2 ALL PROTECTIVE EQUIPMENT SHOULD BE TESTED

This Section lists some protective equipment that has often been overlooked and not included in testing schedules.

14.2.1 HIRED EQUIPMENT

After a low-temperature trip on a nitrogen vaporizer failed to operate it was found that the trip was never tested. The equipment was hired and it was assumed—wrongly—that the leasor would test it.

14.2.2 EMERGENCY VALVES

A pump leaked and caught fire. It was impossible to reach the suction and delivery valves. But there was a second valve in the suction line between the pump and the tank from which it was taking suction, situated in the tank bund. Unfortunately this valve was rarely used and was too stiff to operate.

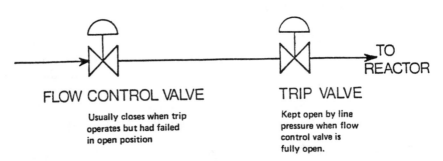

FLOW CONTROL VALVE

Usually closes when trip
operates but had failed
in open position

TRIP VALVE

Kept open by line
pressure when flow
control valve is
fully open.

Figure 14.1. When the control valve was open, the pressure prevented the trip valve closing.

All valves—whether manual or automatic—that may have to be operated in an emergency—should be tested regularly (weekly or monthly). If completely closing a valve will upset production it should be closed halfway and closed fully during shutdowns.

Emergency blowdown valves are among those which should be tested regularly.

14.2.3 STEAM TRACING

A furnace feed pump tripped out. The flowmeter was frozen so the low-flow trip did not operate. Two tubes burst, causing a long and fierce fire. The structure and the other tubes were damaged and the stack collapsed.

In cold weather the trace heating on instruments which form part of trip and alarm systems should be inspected regularly. This can be part of the test routine but more frequent testing may be necessary.

14.2.4 RELIEF VALVES, VENTS, ETC.

Section 10.4.2 listed some items which should be registered for inspection as part of the relief valve register. Section 2.2 (a) described an accident which killed two men. A vent was choked and the end of the vessel was blown off by compressed air.

14.2.5 OTHER EQUIPMENT

Other equipment which should be tested regularly includes:

1. Spare pumps, particularly those fitted with auto-starts
2. Diesel generators
3. Fire and smoke detectors and automatic fire-fighting equipment
4. Water sprays
5. Mechanical protective equipment such as overspeed trips
6. Earth connections, especially the moveable connections used for grounding tank trucks.

The following incidents demonstrate the need to test all protective equipment:

(a) A compressor was started up with the barring gear engaged. The barring gear was damaged.

The compressor was fitted with a protective system that should make it impossible to start the machine with the barring gear engaged. But the protective system was out of order. It was not tested regularly.

(b) In an automatic fire-fighting system a small explosive charge cut a rupture disc and released the fire-fighting agent, Halon. The manufacturers said it was not necessary to test it. To do so, a charge of Halon, which is expensive, would have to be discharged.

The client insisted on a test. The smoke detectors worked and the explosive charge operated but the cutter did not cut the rupture disc. The explosive charge could not develop enough pressure because the volume between it and the rupture disc was too great. The volume had been increased as the result of a change in design: installation of a device for discharging the Halon manually.

One sometimes comes across a piece of protective equipment which it is impossible to test. All protective equipment should be designed so that it can be tested easily.

14.3 TESTING CAN BE OVERDONE

An explosion occurred in a vapor phase hydrocarbon oxidation plant injuring 10 people and seriously damaging the plant, despite the fact that it was fitted with a protective system which measured the oxygen content and isolated the oxygen supply if the concentration approached the flammable limit.

It is usual to install several oxygen analyzers, but this plant was fitted with only one. The management therefore decided to make up for the deficiency in numbers by testing it daily instead of weekly or monthly.

The test took over an hour. The protective system was therefore out of action for about 5% of the time. There was a chance of one in twenty that it would not prevent an explosion because it was being tested. It was, in fact, under test when the oxygen content rose.

14.4 PROTECTIVE SYSTEMS SHOULD NOT RESET THEMSELVES

(a) A gas leak occurred at a plant and caught fire. The operator saw the fire through the window of the control room and operated a switch which should have isolated the feed and opened a blowdown valve. Nothing happened. He operated the switch several times but still nothing happened. He then went outside and closed the feed valve and opened the blowdown valve by hand.

The switch operated a solenoid valve which vented the compressed air line leading to valves in the feed and blowdown lines (Figure 14.2). The feed valve then closed and the blowdown valve

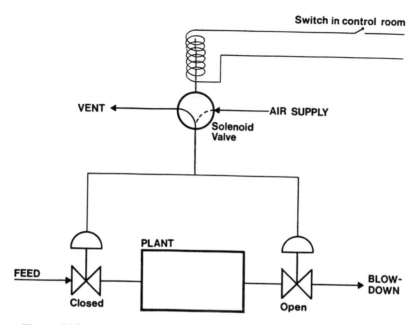

Figure 14.2. An automatic system which will take a minute or so to operate.

opened. This did not happen instantly because it took a minute or so for the air pressure to fall in the relatively long lines between the solenoid valve and the other valves.

The operator expected the system to function as soon as he operated the switch. When it did not, he assumed it was faulty. Unfortunately, after operating the switch several times, he left it in its normal position.

The operator had tested the system on several occasions, as it was used at every shutdown. However, it was tested in conditions of no stress and he did not notice that it took a minute or so to operate.

The solenoid valve should have been fitted with a latch so that once the switch had been operated the solenoid valve could not return to its normal position until it was reset by hand.

(b) A liquid phase hydrocarbon oxidation plant was fitted with a high-temperature trip which shut off the air and opened a drain valve that dumped the contents of the reactor in a safe place (Figure 14.3).

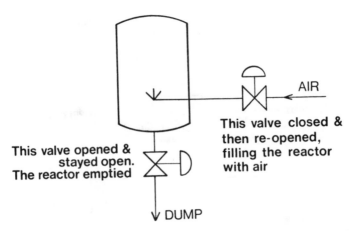

Figure 14.3. When the air valve reopened after a dump, a flammable mixture formed in the reactor.

When the air valve reopened after a dump, a flammable mixture could form in the reactor.

One day the temperature measuring device gave a false indication of high temperature. The air valve closed and the drain valve opened. The temperature indication fell, perhaps because the reactor was now empty. The drain valve stayed open but the air valve reopened and a flammable mixture was formed in the reactor. Fortunately it did not ignite.

The air valve reopened because the solenoid valve in the instrument air line leading to the air valve would not stay in the tripped position. It should have been fitted with a latch.

14.5 TRIPS SHOULD NOT BE DISARMED WITHOUT AUTHORIZATION

Many accidents have occurred because operators made trips inoperative. The following incidents are typical:

(a) Experience shows that when autoclaves or other batch reactors are fitted with drain valves, they may be opened at the wrong time and the contents tipped onto the floor, often inside a building.

To prevent this, the drain valves on a set of reactors were fitted with interlocks so that they could not be opened when the pressure was above a preset value.

Nevertheless, a drain valve was opened when a reactor was up to pressure and a batch emptied onto the floor. The inquiry disclosed that the pressure measuring instruments were not very reliable. So the operators had developed the practice of defeating the interlocks either by altering the indicated pressure with the zero adjustment screw or by isolating the instrument air supply.

One day the inevitable happened. Having defeated the interlock, an operator opened a drain valve by mistake instead of a transfer valve.

Protective equipment may have to be defeated from time to time but this should only be done after authorization in writing by a responsible person. And the fact that the equipment is out of action should be clearly signaled—for example, by a light on the panel.

(b) Soon after a startup, part of a unit was found to be too hot. Flanged joints were fuming. It was then found that the combined temperature controller and high temperature trip had been unplugged from the power supply.

Trips should normally be designed so that they operate if the power supply is lost. If this will cause a dangerous upset in plant operation, then an alarm should sound when power is lost.

Trips should be tested at startup if they have been worked on during a shutdown. Particularly important trips such as those on furnaces and compressors, and high oxygen concentration trips, should always be tested after a major shutdown.

The most common cause of a high temperature (or pressure, flow, level, etc.) is a fault in the temperature measuring or control system.

14.6 INSTRUMENTS SHOULD MEASURE DIRECTLY WHAT WE NEED TO KNOW

An ethylene oxide plant tripped and a light on the panel told the operator that the oxygen valve had closed. Because the plant was going to be restarted immediately, he did not close the hand-operated isolation valve as well. Before the plant could be restarted an explosion occurred. The oxygen valve had not closed and oxygen continued to enter the plant (Figure 14.4).

The oxygen valve was closed by venting the air supply to the valve diaphragm, by means of a solenoid valve. The light on the panel merely

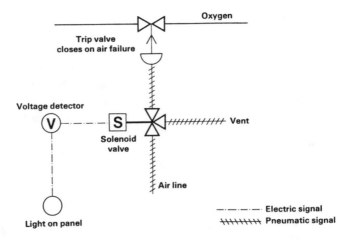

Figure 14.4. The light shows that the solenoid is de-energized, not that the oxygen flow has stopped.

said that the solenoid had been de-energized. Even though the solenoid is de-energized the oxygen flow could have continued because:

1. The solenoid valve did not open.
2. The air was not vented.
3. The trip valve did not close.

Actually the air was not vented. The 1-inch vent line on the air supply was choked by a wasp's nest.

Whenever possible we should measure directly what we need to know and not some other parameter from which it can be inferred (1).

Other incidents in which operators relied on automatic valves and did not back them up with hand valves are described in Sections 17.3 (b) and 17.5 (c).

14.7 TRIPS ARE FOR EMERGENCIES, NOT FOR ROUTINE USE

(a) Section 5.1.1 described how a small tank was filled every day with sufficient raw material to last until the following day. The operator watched the level in the tank and switched off the filling pump when the tank was 90% full. This system worked satisfactorily for several years before the inevitable happened and the operator allowed the tank to overfill. A high level trip was then installed to switch off the pump automatically if the level exceeded 90%. To everyone's surprise the tank overflowed again after about a year.

When the trip was installed it was assumed that:

1. The operator will occasionally forget to switch off the pump in time, and the trip will then operate.
2. The trip will fail occasionally (about once in two years).
3. The chance that both will occur at the same time is negligible.

However, it did not work out like this. The operator decided to rely on the trip and stopped watching the level. The manager and foreman knew this but were pleased that the operator's time was being utilized better. A simple trip fails about once every two years so the tank was bound to overflow after a year or two. The trip was being used as a process controller and not as an emergency instrument.

After the second spillage the following options were considered:

1. Persuade the operator to continue to watch the level. This was considered impracticable if the trip was installed.
2. Remove the trip, rely on the operator and accept an occasional spillage.
3. Install two trips, one to act as a process controller and the other to take over if the first one fails.

(b) When a furnace fitted with a low flow trip has to be shut down, it is common practice to stop the flow and let the low flow trip isolate the fuel supply to the burners. In this way the trip is tested without upsetting production.

On one occasion the trip failed to operate and the furnace coils were overheated. The operator was busy elsewhere on the unit and was not watching the furnace.

All trips fail occasionally. So if we are deliberately going to wait for a trip to operate we should watch the readings and leave ourselves time to intervene if the trip fails to work.

14.8 TESTS MAY FIND FAULTS

Whenever we carry out a test, we may find a fault, and we must be prepared for one.

After changing a chlorine cylinder, two workers opened the valves to make sure there were no leaks on the connecting pipework. They did not expect to find any so they did not wear breathing apparatus. Unfortunately there were some small leaks and they were affected by the chlorine.

The workers' actions were not very logical. If they were sure there were no leaks, there was no need to test. If there was a need to test, then leaks were possible and breathing apparatus should have been worn.

Similarly, pressure tests (at pressures above design, as distinct from leak tests at design pressure) are intended to detect defects. Defects may be present—if we were sure there were no defects we would not need to pressure test—and therefore we must take suitable precautions. No one should be in a position where he may be injured if the vessel or pipework fails.

14.9 SOME MISCELLANEOUS INCIDENTS

(a) A radioactive level indicator on the base of a distillation column was indicating a low level although there was no doubt that the level was normal.

Radiography of pipewelds was in operation 60 m away and the radiation source was pointing in the direction of the radiation detector on the column. When the level in the column is high the liquid absorbs radiation; when the level is low more radiation falls on the detector. The detector could not distinguish between radiation from the normal source and radiation from the radiographic source and registered a low level.

(b) As pointed out in Section 1.5.4 (d) on several occasions fitters have removed thermowells—pockets into which a temperature measuring device is inserted—without realizing that this would result in a leak.

(c) Section 9.2.1 (c) described an incident in which a float came loose from a level controller in a sphere containing propane and formed a perfect fit in the short pipe below the relief valve. When the sphere was filled completely and isolated, thermal expansion caused the 14 m diameter sphere to increase in diameter by 0.15 m (6 inches).

REFERENCE

1. W. H. Doyle, *Some major instrument connected CPI losses,* Chemical Process Industry Symposium, Philadelphia, Pa, 1972.

Chapter 15

Static Electricity

Static electricity (static for short) has been blamed for many fires and explosions, sometimes correctly. Sometimes, however, the investigator has failed to find any other source of ignition. So he assumes that it must have been static even though he is unable to show precisely how a static charge could have been formed and discharged.

A static charge is formed whenever two surfaces are in relative motion, for example, when a liquid flows past the walls of a pipeline, when liquid droplets or solid particles move through the air or when a man walks, gets up from a seat or removes an article of clothing. One charge is formed on one surface—for example, the pipe wall—and an equal and opposite charge on the other surface—for example, the liquid flowing past it.

Many static charges flow rapidly to earth as soon as they are formed. But if a charge is formed on a nonconductor or on a conductor which is not grounded, they can remain for some time. If the level of the charge, the voltage, is high enough the static will discharge by means of a spark which can ignite any flammable vapors which may be present. Examples of nonconductors are plastics and nonconducting liquids such as most pure hydrocarbons. Most liquids containing oxygen atoms in the molecule are good conductors.

Even if a static spark ignites a mixture of flammable vapor and air, it is not really correct to say that static electricity was the "cause" of the fire or explosion. The real cause was the leak or whatever event led to the formation of a flammable mixture. Once flammable mixtures are formed, experience shows that sources of ignition are likely to turn up. The deliberate formation of flammable mixtures should never be allowed except when the risk of ignition is accepted—for example, in the vapor spaces of fixed roof tanks containing flammable nonhydrocarbons. See Section 5.4.

15.1 STATIC ELECTRICITY FROM FLOWING LIQUIDS

Section 5.4.1 describes explosions in storage tanks, and Section 13.3 describes explosions in tank trucks, ignited by static sparks. The static was formed by the flow of a nonconducting liquid and the spark discharges occurred *between the body of the liquid and the grounded metal containers* (or filling arms).

If a conducting liquid such as acetone or methanol flows into an ungrounded metal container, the container acquires a charge from the liquid and a spark may occur *between the container and any grounded metal that is nearby,* as in the following incidents.

(a) Acetone was regularly drained into a metal bucket. One day the operator hung the bucket on the drain valve instead of placing it on the metal surface below the valve (Figure 15.1).

  The handle of the bucket was covered with plastic. When acetone was drained into the bucket, a static charge accumulated on the acetone and on the bucket. The plastic prevented the charge flowing to earth via the drain pipe, which was grounded. Finally a spark passed between the bucket and the drain valve and the acetone caught fire.

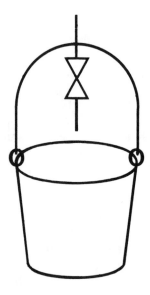

Figure 15.1. The bucket was not earthed and acquired a charge.

Even if the bucket had been grounded it would still have been bad practice to handle a flammable (or toxic or corrosive) liquid in an open container. It should have been handled in a closed can in order to prevent spillages. See Sections 7.1.3 and 12.2 (c). Closed cans, however, will not prevent ignition by static electricity as the following incidents show.

(b) A man held a 10-liter metal container while it was filled with acetone. When he tried to close the valve in the acetone line the acetone ignited and the fire spread to other parts of the building. The man was wearing insulating (crepe rubber) shoes and it is believed that a static charge accumulated on the acetone, the can and the man. When he put his hand near the valve a spark jumped from him to the valve, which was grounded, and ignited the acetone vapor.

(c) Metal drums were occasionally filled with vinyl acetate via a 2-inch-diameter rubber hose. There was no means for grounding the drum and the rubber hose did not reach to the bottom of the drum; the liquid splashed down from a height of 0.6 m. A few minutes after filling started there was a violent explosion and the ends of the drum were blown out. One end hit a man in the legs, breaking both of them, and the other end broke another man's ankle. He was burned in the ensuing fire and died a few days later.

Note that as in the incident described in Section 13.3 the operation had been carried out a number of times before conditions were right for an explosion to occur.

15.2 STATIC ELECTRICITY FROM GAS AND WATER JETS

On a number of occasions people have received a mild electric shock while using a carbon dioxide fire extinguisher. The gas jets from the extinguishers contain small particles of solid carbon dioxide so a charge will collect on the horn of the extinguisher and may pass to earth via the hand of the person who is holding the horn.

A more serious incident of the same sort occurred when carbon dioxide was used to inert the tanks of a ship which had contained naphtha. An explosion occurred, killing four men and injuring seven. The carbon dioxide was added through a plastic hose 8 m long which ended in a short brass hose (0.6 m long) that was dangled through the ullage hole of one of the tanks. It is believed that a charge accumulated on the brass hose and a spark passed between it and the tank (1).

A few years later carbon dioxide was injected into an underground tank containing jet fuel as a tryout of a fire-fighting system. The tank blew up, killing 18 people who were standing on top of the tank. In this case the discharge may have occurred from the cloud of carbon dioxide particles.

The water droplets from steam jets are normally charged and discharges sometimes occur from the jets to neighboring grounded pipes. These discharges are of the corona type rather than true sparks and may be visible at night; they look like a small flame (2).

Discharges from water droplets in ships' tanks (being cleaned by high pressure water washing equipment) have ignited flammable mixtures and caused serious damage to several supertankers (3). The discharges occurred from the cloud of water droplets, and were thus "internal lightning."

A glass distillation column cracked and water was sprayed onto the crack. A spark was seen to jump from the metal cladding on the insulation, which was not grounded, to the end of the water line. Although no ignition occurred in this case, the incident shows the need to ground all metal objects and equipment. They may act as collectors for charges from steam leaks or steam or water jets.

Most equipment is grounded by connection to the structure or electric motors. But this may not be true of insulation cladding, scaffolding, pieces of scrap or tools left lying around, or pieces of pipe attached by nonconducting pipe or hose (see next item). In one case, sparks were seen passing from the end of a disused instrument cable; the other end of the cable was exposed to a steam leak.

15.3 STATIC ELECTRICITY FROM POWDERS

A powder was emptied down a metal duct into a plant vessel. The duct was replaced by a rubber hose as shown in Figure 15.2.

The flow of powder down the hose caused a charge to collect on it. Although the hose was reinforced with metal wire and was therefore conducting, it was connected to the plant at each end by short polypropylene pipes which were nonconducting. A charge therefore accumulated on the hose, a spark occurred, the dust exploded and a man was killed.

A nonconducting hose would have held a charge. But a spark from it would not have been as big as from a conducting hose and might not have ignited the dust, though we cannot be certain. It would have been safer than an ungrounded conducting hose but less safe than a grounded conducting hose.

Hoses and ducts used for conveying explosive powders should be made from conducting material and be grounded throughout. Alternatively (or additionally) the atmosphere can be inerted with nitrogen, the ducts can be made strong enough to withstand the explosion or an explosion vent can be provided.

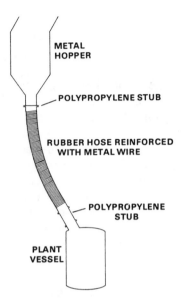

Figure 15.2. The flow of powder caused a static charge to collect on the insulated hose.

This incident is another example of a hazard introduced by a simple plant modification. Other incidents are described in Chapter 2.

Note that introducing a plastic section in a pipeline so that the metal pipe beyond it is no longer grounded can be a hazard with liquids as well as powders.

15.4 STATIC ELECTRICITY FROM CLOTHING

(a) An operator slipped on a staircase, twisted his ankle and was absent for 17 shifts. The staircase was in good condition and so were the operator's boots.

Many people's reaction would have been that this is another of those accidents that we can do nothing about, another occasion when "Man told to take more care" appears on the accident report.

However, in the plant where the accident occurred they were not satisfied with this easy way out. They looked into the accident more thoroughly. The injured man was asked why he had not used the handrails.

It then came to light that the handrails were covered with plastic and that anyone using them *and wearing insulating footware* acquired an electric charge. When he touched the metal of the plant he got a mild electric shock. The spark, of course, was not serious enough to cause any injury. But it was unpleasant. People therefore tended not to use the handrails.

For a spark to be felt it must have an energy of a least 1 mJ. The minimum energy required to ignite a flammable mixture is 0.2 mJ so a spark that can be felt is certainly capable of causing ignition, if flammable vapor is present.

(b) We have all acquired a static charge by walking across a man-made fiber carpet (or just by getting up from our chairs) and then felt a mild shock when we touched a metal object such as a filing cabinet. Similar charges can be acquired by walking across a plant floor wearing nonconducting footware. And sparks formed in this way have been known to ignite leaks of flammable gas or vapor, especially in dry climates. However the phenomenon is rare. It does not normally justify insistence on the use of conducting footware.

(c) A driver drew up at a filling station, removed the cap from the end of the filler pipe and held it in his hand while an attendant filled the car with gasoline. The driver took off his pullover, thus acquiring a charge and leaving an equal and opposite charge on the pullover which he threw into the car. He was wearing nonconducting shoes so the charge could not leak away to earth.

When he was about to replace the cap on the end of the filler pipe a spark jumped from the cap to the pipe and a flame appeared on the end of the pipe. It was soon extinguished. The flame could not travel back into the gasoline tank. The mixture of vapor and air in the tank was too rich to explode.

At one time there was concern that man-made fiber clothing might be more likely than wool or cotton clothing to produce a charge on the wearer. The incident just described shows that a charge is acquired only when the clothing is removed. When dealing with a leak we do not normally start by removing our clothing. There is therefore no need to restrict the types of cloth used, so far as static electricity is concerned.

REFERENCES

1. *Fire Journal,* Nov. 1967, p. 89.
2. A. F. Anderson, *Electronics and Power,* January 1978.
3. *S. S. Mactra (ON 337004)—Report of Court No. 8057 Formal Investigation,* Her Majesty's Stationery Office, London, 1973.

Chapter 16

Materials of Construction

16.1 WRONG MATERIAL USED

Many incidents have occurred because the wrong material of construction was used. This has usually been the result of errors by maintenance or construction personnel or suppliers, who did not use or did not supply the materials specified. Few failures have been the result of errors by materials specialists who incorrectly specified the materials to be used.

The following incidents are typical:

(a) A titanium flange was fitted by mistake on a line carrying dry chlorine. The flange caught fire. Titanium is ideal for wet chlorine but catches fire on contact with dry chlorine. (Burning in this case means rapid combination with chlorine, not oxygen.)

(b) A carbon steel valve painted with aluminum paint was used instead of a stainless steel valve. It corroded rapidly.

(c) A plug valve was supplied with a pure nickel plug instead of one made from 304L stainless steel. The valve body was made from the correct material. The valve was installed in a nitric acid line. Five hours later the plug had disappeared and acid was escaping through the stem seal.

The manufacturers had provided a test certificate stating that the valve was made from 304L steel.

(d) During the night a valve had to be changed on a unit which handled a mixture of acids. The fitter could not find a suitable valve in the workshop but on looking around he found one on another unit. He tested it with a magnet and finding it to be nonmagnetic he assumed

it was similar to the stainless steel valves normally used. He therefore installed it.

Four days later the valve was badly corroded and there was a
spillage of acids.

The valve was made of Hastalloy, an alloy suitable for use on the
unit where it was found but not suitable for use with the mixture of
acids on the unit on which it had been installed.

(e) A tank truck, used for internal transport, looked as if it was made
of stainless steel. It was therefore filled with 50% caustic soda solution. Twelve hours later the tank was empty. It was made of aluminum and the caustic soda created a hole and leaked out.

The material of construction has now been stenciled on all tank
trucks used for internal transport in the plant where the incident
occurred.

(f) A small new tank was installed with an unused branch blanked off.
A month later the branch was leaking. It was then discovered that
the tank had arrived with the branch protected by a blank flange
made of wood. The wood was painted the same color as the tank
and nobody realized that it was not a steel blank.

(g) A leak on a refinery pump which was followed by a fire was due to
incorrect hardness of the bolts used. Other pumps supplied by the
same manufacturer were then checked and another was found with
off-specification bolts. The pump had operated for 6,500 hours before the leak occurred.

If the pump had been fitted with a remotely operated emergency
isolation valve as recommended in Section 7.2.1, the leak could
have been stopped quickly. Damage would have been slight. As it
was the unit was shut down for five weeks.

(h) Section 9.1.6 (b) described what happened when the exit pipe of a
high-pressure ammonia converter was made from carbon steel instead of $1\frac{1}{4}\%$ Cr, 0.5% Mo. Hydrogen attack occurred, a hole appeared at a bend and the reaction forces from the escaping gas
pushed the converter over.

(i) After some new pipes were found to be made of the wrong alloy,
further investigation showed that many of the pipes, clips and
valves in store were made of the wrong alloys. The investigation
was extended to the rest of the plant and the following are some
examples of the findings:

1. The wrong electrodes had been used for 72 welds on the tubes of
a fired heater.
2. Carbon steel vent and drain valves had been fitted on an alloy
steel system.

3. An alloy steel heat exchanger shell had been fitted with two large carbon steel flanges. The flanges were stamped as alloy.

(j) Checks carried out on the materials delivered for a new ammonia plant showed that 5,480 items (1.8% of the total) were delivered in the wrong material. These included 2,750 furnace roof hangers; if the errors had not been spotted the roof would probably have failed in service.

> ". . . . vendors often sent without notice what they regarded as 'superior' material. Thus, if asked to supply 20 flanges in carbon steel of a given size, the vendor, if he had only 19 such flanges available, was quite likely to add a 20th of the specified size in 'superior' $2\frac{1}{4}\%$ Cr. When challenged the vendor was often very indignant because he had supplied 'superior,' i.e., more expensive, material at the original price. We had to explain that the 'superior' material was itself quite suitable, *if we knew about it*. If we didn't, we were quite likely to apply the welding procedures of carbon steel to $2\frac{1}{4}\%$ Cr steel with unfortunate results (1)."

As the result of incidents such as those described in (c) and (g) through (j) above, many companies now insist that if the use of the wrong grade of steel can affect the integrity of the plant, all steel (flanges, bolts, welding rods, etc. as well as pipes) must be checked for composition before use. The analysis can be carried out easily with a spectrographic analyzer.

The design department should identify which pipelines, etc., need to be checked and should mark drawings accordingly.

(k) The recycle of scrap to produce stainless steel has led to increases in the concentration of trace elements not covered in the steel specifications. This may lead to poorer corrosion resistance and weld quality, though so far no dangerous incidents have been reported.

16.2 HYDROGEN PRODUCED BY CORROSION

Hydrogen produced by corrosion can turn up in unexpected places as shown by the following incidents:

(a) An explosion occurred in a tank containing sulphuric acid. As the possibility of an explosion had not been foreseen, the roof/wall weld was stronger than usual and the tank split at the base/wall weld. The tank rose 15 m into the air, went through the roof of the building and fell into an empty piece of ground close by, just missing other tanks. Fortunately no one was hurt. If the tank had fallen on the other side of the building it would have fallen into a busy street.

Slight corrosion in the tank had produced some hydrogen. The tank was fitted with an overflow pipe leading down to the ground, but no vent. So the hydrogen could not escape and it accumulated under the conical roof. The hydrogen was ignited by welders working nearby. (Presumably some found its way out of the overflow.) (2).

The tank should have been fitted with a vent at the highest point as shown in Figure 16.1.

Many suppliers of sulphuric acid recommend that it is stored in pressure vessels designed to withstand a gauge pressure of 30 psi (2 bar). The acid is usually discharged from tank trucks by compressed air and if the vent is choked the vessel could be subjected to the full pressure of the compressed air.

(b) Hydrogen produced by corrosion is formed as atomic hydrogen. It can diffuse through iron. This has caused hydrogen to turn up in unexpected places such as the insides of hollow pistons. When holes have been drilled in the pistons, the hydrogen has come out and caught fire (3).

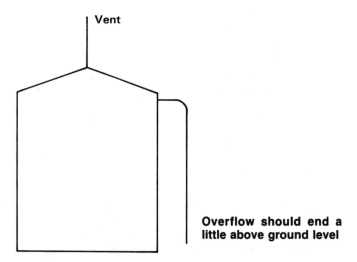

Figure 16.1. Acid tanks should be fitted with a high-point vent as well as an overflow so that hydrogen can escape.

In another case acidic water was used to clean the inside of the water jacket which surrounded a glass-lined vessel. Some hydrogen diffused through the wall of the vessel and developed sufficient pressure to crack the glass lining.

16.3 OTHER EFFECTS OF CORROSION

Corrosion usually results in a leak or failure of a support because a vessel or support gets too thin. It is then not strong enough to withstand the pressure or load. However, rust can cause failure in another way. It occupies about seven times the volume of the steel from which it was formed. When rust occurs between two plates that have been bolted or riveted together, a high pressure develops. This can force the plates apart or even break the bolts or rivets.

REFERENCES

1. G. C. Vincent and C. W. Gent, *Ammonia Plant Safety,* Volume 20, 1978, p. 22.
2. *Chemical Safety Summary,* London, Chemical Industries Association, No. 192, October/December 1977.
3. *Case Histories of Accidents in the Chemical Industry, No. 1807,* Manufacturing Chemists Association, April 1975.

Chapter 17

Operating Methods

This chapter describes some accidents that occurred because of defects in operating procedures. It does not include accidents which occurred because of defects in procedures for preparing equipment for maintenance or vessels for entry. These are discussed in Chapters 1 and 11.

17.1 TRAPPED PRESSURE

Trapped pressure is a familiar hazard in maintenance operations and is discussed in Section 1.3.6. Here we discuss accidents which have occurred as a result of process operation.

Every day, in every plant, equipment which has been under pressure is opened up. This is normally done under a work permit. One man prepares the job and another opens up the vessel. And it is normally done by slackening bolts so that any pressure present will be detected before it can cause any damage—provided the joint is broken in the correct way, described in Section 1.5.1.

Several fatal or serious accidents have occurred when one man has carried out the whole job—preparation and opening up—and has used a quick-release fitting instead of nuts and bolts. One incident, involving a tank truck, was described in Section 13.5. Here is another:

A suspended catalyst was removed from a process stream in a pressure filter. After filtration was complete, the remaining liquid was blown out of the filter with steam at a gauge pressure of 30 psi (2 bar). The pressure in the filter was blown off through a vent valve and the fall in pressure was observed on a pressure gauge. The operator then opened the filter for cleaning. The filter door was held closed by eight radial bars which fitted into U-bolts on the filter body. The bars were withdrawn from the U-bolts by turning a large wheel, fixed to the door. The door could then be withdrawn.

One day an operator started to open the door before blowing off the pressure. As soon as he opened it a little it blew open and he was crushed between the door and part of the structure and was killed instantly.

In situations such as this it is inevitable that sooner or later an operator will forget that he has not blown off the pressure and will attempt to open up the equipment while it is still under pressure. On this occasion the operator was at the end of his last shift before starting his vacation.

As with the accidents described in Section 3.2, it is too simple to say that the accident was due to the operator's mistake. The accident was the result of a situation that made it almost inevitable.

Whenever an operator has to open up equipment which has been under pressure:

(a) The design of the door or cover should allow it to be opened about 1/4 inch while still capable of carrying the full pressure and a separate operation should be required to release the cover fully. If the cover is released while the vessel is under pressure, then this is immediately apparent and the pressure can blow off through the gap or the cover can be resealed. And in addition, if possible,

(b) interlocks should be provided so that the vessel cannot be opened up until the source of pressure is isolated and the vent valve is open.

(c) the pressure gauge and vent valve should be visible to the operator when he is about to open the door or cover (1).

17.2 CLEARING CHOKED LINES

(a) A man was rodding out a choked 1/4-inch line leading to an instrument [Figure 17.1 (a)]. When he had cleared the choke he found that the valve would not close and he could not stop the flow of flammable liquid. Part of the unit had to be shut down.

Rodding out *narrow bore* lines is sometimes necessary. But before doing so a ball valve or cock should be fitted on the end [Figure 17.1 (b)]. It is then possible to isolate the flow when the choke has been cleared even if the original valve will not close.

(b) Compressed air at a gauge pressure of 50 psi (3.4 bar) was used to clear a choke in a 2-inch line. The solid plug got pushed along with such force that when it reached a slip-plate (spade), the slip-plate was knocked out of shape, rather like the one shown in Figure 1.15.

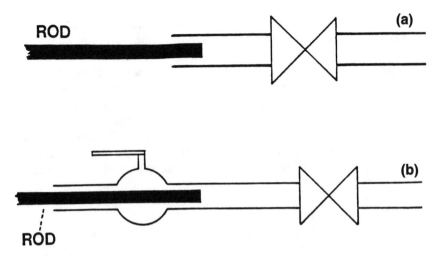

Figure 17.1. The wrong (a) and right (b) ways of clearing a choked line.

On another occasion a 4-inch-diameter vertical U-tube, part of a large heat exchanger, was being cleaned mechanically when the cleaning tool, weight about 25 kg, stuck in the tube. A supply of nitrogen at a gauge pressure of 3,000 psi (200 bar) was available so it was decided to use it to try to clear the choke. The tool shot out of the end of the U-tube and came down through the roof of a building 100 m away.

Gas pressure should never be used for clearing choked lines.

17.3 FAULTY VALVE POSITIONING

Many accidents have occurred because operators failed to open (or close) valves when they should have done so. Most of these incidents occurred because operators forgot to do so and such incidents are described in Sections 3.2.7, 3.2.8, 13.5 and 17.1. In this section we discuss incidents which occurred because operators did not understand why valves should be open (or closed).

(a) As described in Section 3.3.4, the emergency blowdown valves on a plant were kept closed by a hydraulic oil supply. One day the valves opened and the plant started to blow down. It was then discovered that, unknown to the manager, the foreman had developed the practice, contrary to instructions, of isolating the oil supply valve "in case the supply pressure in the oil system failed." This was a most unlikely occurrence and much less likely than the oil pressure leaking away from an isolated system.

(b) The air inlet to a liquid phase oxidation plant became choked from time to time. To clear the choke the flow of air was isolated and some of the liquid in the reactor was allowed to flow backwards through the air inlet and out through a purge line which was provided for this purpose (Figure 17.2).

One day the operator closed the remotely operated valve in the air line but did not consider it necessary to close the hand valve as well, although the instructions said that he should. The remotely-operated valve was leaking, the air met the reactor contents in the feed line and reaction took place there. The heat developed caused the line to fail and a major fire followed.

The air line should have been provided with remotely-operated double block and bleed valves, operated by a single button.

Other incidents in which operators relied on automatic valves and did not back them up with hand valves are described in Sections 14.6 and 17.5 (c).

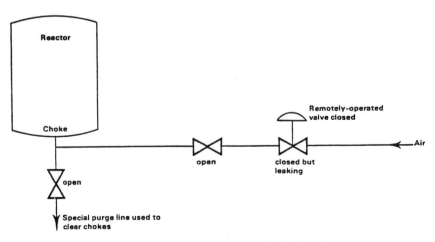

Figure 17.2. Liquid purge burned in the drain line.

17.4 RESPONSIBILITIES NOT DEFINED

The following incident shows what can happen when responsibility for plant equipment is not clearly defined and operators in different teams, responsible to different managers, operate the same valves.

The flare stack shown in Figure 17.3 was used to dispose of surplus fuel gas, which was delivered from the gasholder by a booster through valves B and C. Valve C was normally left open because valve B was more accessible.

One day the operator responsible for the gasholder saw that it had started to fall. He therefore imported some gas from another unit. Nevertheless, a half hour later the gasholder was sucked in.

Another flare stack at a different plant had to be taken out of service for repair. An operator at this plant therefore locked open valves A and B so that he could use the "gasholder flarestack." He had done this before, though not recently, and some changes had been made since he last used the flarestack. He did not realize that his action would result in the gasholder emptying itself through valves C and B. He told three other men what he was going to do but he did not tell the gasholder operator. He did not know that this man was concerned.

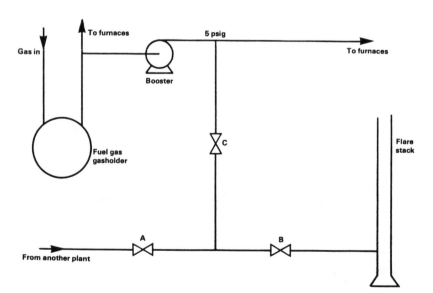

Figure 17.3. Valve B was operated by different operators.

Responsibility for each item of equipment should be clearly defined at manager, foreman, and operator level and only the men responsible for each item should operate it. If different teams are allowed to operate the same equipment then sooner or later an incident will occur.

17.5 COMMUNICATION FAILURES

This section describes some incidents which occurred because of a failure to tell people what they need to know, because of failures to understand what had been told or because of misunderstanding about the meanings of words.

(a) A maintenance foreman was asked to look at a faulty cooling water pump. He decided that, to prevent damage to the machine, it was essential to reduce its speed immediately. He did so, but did not tell any of the operating team straight away. The cooling water rate fell, the process was upset and a leak developed on a cooler.

(b) A tank truck which had contained liquefied petroleum gas was being swept out before being sent for repair. The laboratory staff was asked to analyze the atmosphere in the tanker to see if any hydrocarbon was still present. The laboratory staff regularly analyzed the atmosphere inside LPG tank trucks to see if any oxygen was present. Owing to a misunderstanding they assumed that an oxygen analysis was required on this occasion and reported over the telephone "none detected." The operator assumed that no hydrocarbon had been detected and sent the tank truck for repair.

Fortunately the garage had their own check analysis carried out. This showed that LPG was still present—actually over 1 ton of it.

For many plant control purposes telephone results are adequate. But when analyses are made for safety reasons, results should be accepted only in writing.

(c) A batch vacuum still was put on standby because there were some problems in the unit which took the product. The still boiler was heated by a heat transfer oil and the supply was isolated by closing the control valve. The operators expected that the plant would be back on line soon so they did not close the hand isolation valves and they kept water flowing through the condenser. However, the vacuum was broken and a vent on the boiler was opened.

The problems at the downstream plant took much longer than expected to correct, and the batch still stayed on standby for five days. No readings were taken and when recorder charts ran out they were not replaced.

The heat transfer control valve was leaking. Unknown to the operators the boiler temperature rose from 75°C to 143°C, the boiling point of the contents. Finally, bumping in the boiler caused about 0.2 ton of liquid to be discharged through the vent.

Other incidents which occurred because operators relied on automatic valves and did not back them up with hand valves were described in Sections 14.6 and 17.3 (b). In this incident the point to be emphasized in addition is that the operators were not clear on the difference between a standby and a shutdown. No maximum period for standby was defined. And no readings were taken during periods on standby. Plant instructions should give guidance on both these matters.

(d) Designers often recommend that equipment is "checked" or "inspected" regularly. But what do these words mean? They should state precisely what tests should be carried out and what they hope to determine by them.

In 1961 a brake component in a colliery elevator failed, fortunately without serious consequences. An instruction was issued that all similar components should be examined. It did not say how or how often. At one colliery the component was examined in position but was not removed for complete examination and was not scheduled for regular examination in the future.

In 1973 it failed and 18 men were killed (2).

(e) Under the UK Ionizing Radiation (Sealed Sources) Regulations all sealed radioactive sources must be checked by an authorized person "each working day" to make sure that they are still in position.

Following an incident at one plant it was found that they took this to mean that the authorized person must check the presence of the sources on Mondays to Fridays, but not on weekends.

However, "each working day" means each day that the radioactive source is working, not each day that the authorized person is working!

REFERENCES

1. T. A. Kletz, *Loss Prevention,* Vol. 13, 1980, p. 1.
2. *Accident at Markham Colliery, Derbyshire,* London, Her Majesty's Stationery Office, 1974.

Chapter 18

Reverse Flow and Other Unforeseen Deviations

This chapter describes some incidents which occurred because of deviations from the design intention, as expressed in the process flow diagrams (also known as line diagrams or process and instrumentation diagrams). The fact that these deviations could occur was not spotted during the design stage and they had unfortunate unforeseen results. Ways of spotting these deviations by hazard and operability studies are discussed in Section 18.7.

Errors introduced during modifications are discussed in Chapter 2, while designs which provided opportunities for operators to make errors were discussed in Chapter 3.

One of the commonest errors made at the process-flow-diagram stage is failure to foresee that flow may take place in the reverse direction to that intended, as discussed next (1).

18.1 REVERSE FLOW FROM A PRODUCT RECEIVER OR BLOWDOWN LINE BACK INTO THE PLANT

(a) Accidents have occurred because gas flowed from a product receiver into a plant which was shut down and depressured. In one incident ammonia flowed backward from a storage vessel, through a leaking valve, into a reflux drum, into a still and out of an open end in the bottoms line, which was open for maintenance (Figure 18.1).

If the possibility of reverse flow had been foreseen then a slip-plate could have been inserted in the line leading to the ammonia storage vessel, as described in Section 1.1.

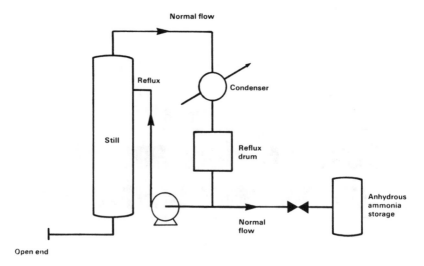

Figure 18.1. Reverse flow occurred from the storage vessel to the open end.

(b) In another incident a toxic gas in a blowdown header flowed through a leaking blowdown valve into a tower and out of the drain valve. The operator who was draining the tower was killed (Figure 18.2).

18.2 REVERSE FLOW INTO SERVICE MAINS

This occurs when the pressure in the service line is lower than usual or when the pressure in the process line is higher than usual. Many plants have experienced incidents such as the following:

1. A steam line had ice on the outside after it had been blown down and liquefied gas had leaked into it.
2. A leak on a nitrogen line caught fire.
3. The paint was dissolved in a cabinet that was pressurized with nitrogen; acetone had leaked into the nitrogen (2).
4. A compressed air line was choked with phenol.
5. Toxic fumes in a steam system affected a man who was working on the system (See Section 1.1.4).

Another incident is illustrated in Figure 18.3. Town-water should never be directly connected to process lines by a hose or permanent connections. A break tank should be provided.

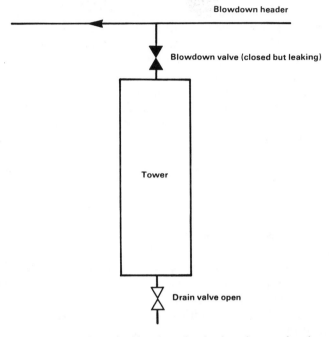

Figure 18.2. Flow occurred from the blowdown header into the vessel and out of the drain valve.

The tea tasted funny...

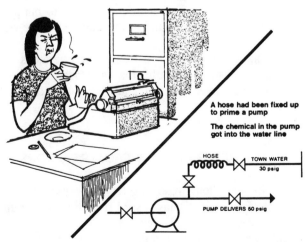

Figure 18.3. Never connect town water to process equipment.

A service which is used intermittently should be connected to process equipment by a hose which is disconnected when not in use, or by double block and bleed valves. If a hose is used, it should be provided with a vent so that it can be depressured before it is disconnected.

If a service is used continuously, it may be connected permanently to process lines. If the service pressure is liable to fall below the normal process pressure, then a low pressure alarm should be provided on the service line. If the process pressure is liable to rise above the normal service pressure, then a high-pressure alarm should be provided on the process side.

In addition, nonreturn valves should be fitted on the service lines.

18.3 REVERSE FLOW THROUGH PUMPS

If a pump trips (or is shut down and not isolated) it can be driven backward by the pressure in the delivery line and damaged. Nonreturn valves are usually fitted to prevent reverse flow but they sometimes fail.

When the consequences of reverse flow are serious, then the nonreturn valve should be scheduled for regular inspection. The use of two, preferably of different types, in series, should be considered. The use of reverse rotation locks should also be considered.

In one plant light oil was pumped at intervals from a tank at atmospheric pressure to one at a gauge pressure of 15 psi (1 bar). For many years the practice was not to close any isolation valves but to rely on the nonreturn valve in the pump delivery. One day a piece of wire got stuck in the nonreturn valve, oil flowed backward and the atmospheric tank overflowed (Figure 18.4).

This is a good example of an accident waiting to happen. Sooner or later the nonreturn valve was bound to fail and a spillage was then inevitable.

In this case the design was not at fault. The operators did not understand the design philosophy. Would this have been foreseen in a hazard and operability study (Section 18.7) and special attention paid to the point in operator training?

18.4 REVERSE FLOW FROM REACTORS

The most serious incidents resulting from reverse flow have occurred when reactant A (Figure 18.5) has passed from the reactor up the reactant B feed line and reacted violently with B.

In one incident paraffin wax and chlorine were reacted at atmospheric pressure. Some paraffin traveled from the reactor back up the chlorine

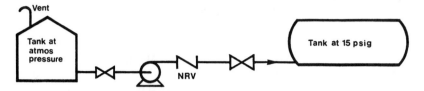

Figure 18.4. The nonreturn valve was relied on to prevent back flow and the isolation valves were not used. A spillage was inevitable.

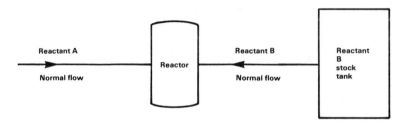

Figure 18.5. Reverse flow of A occurred from the reactor into the B stock tank.

line and reacted with liquid chlorine in a catchpot which exploded with great violence. Bits were found 30 m away (3).

A more serious incident occurred at a plant in which ethylene oxide and aqueous ammonia were reacted to produce ethanolamine. Some ammonia got back into the ethylene oxide storage tank, past several nonreturn valves in series and a positive pump. It got past the pump through the relief valve which discharged into the pump suction line. The ammonia reacted with 30 m³ of ethylene oxide in the storage tank. There was a violent rupture of the tank followed by an explosion of the vapor cloud which caused damage and destruction over a wide area (4).

When such violent reactions can occur it is not sufficient to rely on nonreturn valves. In addition, either:

1. The reactant(s) should be added via a small break tank so that if reverse flow occurs only a small quantity will react and not the main stock, or:
2. The pressure drop in the pipeline should be measured and if it gets too low a trip valve should be closed automatically. A very reliable, duplicated system may be necessary (5).

18.5 REVERSE FLOW FROM DRAINS

This has often caused flammable liquids to turn up in some unexpected places. For example, construction had to be carried out next to a compound of small tanks. Sparks would fall onto the compound. Therefore all flammable liquids were removed from the tanks while the construction took place. Nevertheless a small fire occurred in the compound.

Water was being drained from a tank on another part of the plant. The water flow was too great for the capacity of the drains so the water backed up into the compound of small tanks, taking some light oil with it. This oil was ignited by welding sparks.

18.6 FORWARD FLOW GREATER THAN FORESEEN

Figure 18.6 shows part of an old unit. Valve A could pass a higher rate than valve B. Inevitably, in the end the lower tank overflowed.

18.7 A METHOD FOR FORESEEING DEVIATIONS

The incidents listed earlier in this chapter and many others could have been foreseen if the design had been subjected to a hazard and operability study (HAZOP). This is a technique which allows people to let their imaginations go free and think of all possible ways in which hazards or

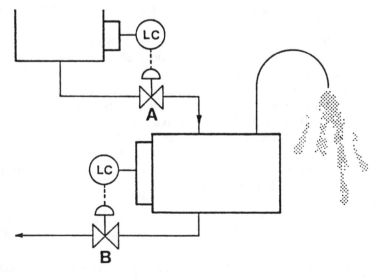

Figure 18.6. Valve A could pass a bigger rate than valve B, thus making a spillage inevitable.

operating problems might arise. But to reduce the chance that something is missed, HAZOP is done in a systematic way, each pipeline and each sort of hazard being considered in turn.

A pipeline for this purpose is one joining two main plant items—for example, we might start with the line leading from the feed tank through the feed pump, to the first feed heater. A series of guide words are applied to this line in turn, the words being:

NONE
MORE OF
LESS OF
PART OF
MORE THAN
OTHER

NONE, for example, means no forward flow or reverse flow when there should be forward flow. We ask:

Could there be no flow?
If so, how could it arise?
What are the consequences of no flow?
How will the operators know that there is no flow?
Are the consequences hazardous or do they prevent efficient operation?
If so, can we prevent no flow (or protect against the consequences) by changing the design or method of operation?
If so, does the size of the hazard or problem justify the extra expense?

The same questions are then applied to "reverse flow" and we then move on to the next word, MORE OF. Could there be "more flow" than design? If so, how could it arise? And so on. The same questions are asked about "more pressure" and "more temperature," and, if they are important, about other parameters such as "more radioactivity" or "more viscosity?"

The technique is described in References 5–10.

REFERENCES

1. T. A. Kletz, *Hydrocarbon Processing,* Vol. 55, No. 3, March 1976, p. 187.
2. T. A. Kletz, *Hydrocarbon Processing,* Vol. 59, No. 11, Nov. 1979, p. 373.

3. D. B. de Oliveria, *Hydrocarbon Processing,* Vol. 52, No. 3, March 1973, p. 113.
4. J. E. Troyan and R. Y. Le Vine, *Loss Prevention,* Vol. 2, 1968, p. 125.
5. H. G. Lawley, *Hydrocarbon Processing,* Vol. 55, No. 4, April 1976, p. 247.
6. H. G. Lawley, *Chemical Engineering Progress,* Vol. 70, No. 4, April 1974, p. 45.
7. *Hazard and Operability Studies,* Chemical Industries Association, London, 1977.
8. R. E. Knowlton, *An Introduction to Hazard and Operability Studies,* Chemetics International, Vancouver, 1981.
9. T. A. Kletz, *Hazop and Hazan—Notes on the Identification and Assessment of Hazards,* Institution of Chemical Engineers, 1983.
10. F. P. Lees, *Loss Prevention in the Process Industries,* Butterworths, 1980, Chapter 21.

Epilogue

Bhopal, Mexico City, Sao Paulo

BHOPAL

While this book was in production, the worst disaster in the history of the chemical industry occurred in Bhopal, in the state of Madhya Pradesh in central India, on December 3, 1984. A leak of methyl isocyanate from a chemical plant, where it was used as an intermediate in the manufacture of a pesticide, spread beyond the plant boundary and caused the poisoning death of over 2,500 people—injuring about ten times as many. The worst incidents to have occurred before 1984 were the explosion of a 50:50 mixture of ammonium sulfate and ammonium nitrate at Oppau in Germany in 1921, which killed 430 people, including 50 members of the public, and the explosion of a cargo of ammonium nitrate in a ship in Texas City harbor in 1947, which killed 552 people, mostly members of the public (1).

At the time of this writing the full technical causes of the Bhopal disaster are not known, but nevertheless it is possible to comment on some of the wider issues.

Methyl isocyanate (isocyanatomethane) boils at about 40°C at atmospheric pressure. According to press reports, the contents of the storage tank overheated and boiled, causing the relief valves to lift. The discharge of vapor—about 25 tons—was too great for the capacity of the scrubbing system and the escaping vapor spread beyond the plant boundary where a shanty town had sprung up.

The press reports suggested that the cause of the overheating was contamination of the methyl isocyanate, by water and/or other materials, and several possible mechanisms were suggested. According to some reports cyanide was produced. If contamination did indeed occur, it reinforces the message of Chapter 18, that hazard and operability studies should be carried out on all new plant designs. They provide a very effective means of showing up ways in which contamination can occur.

Press reports also suggested that the scrubbing system, which should have absorbed the vapor discharged from the relief valve, the flare system which should have burned any vapor which got past the scrubbing system, and the system for cooling the tank were not in full working order, perhaps because the unit using the methyl isocyanate was shut down. If any of these statements are true, they reinforce the point made in Chapter 14, that all protective equipment should be properly maintained and tested regularly. It is easy to buy safety equipment—all you need is money. It is much more difficult to see that, year in, year out, it is kept in full working order. Procedures are subject to a form of corrosion more rapid than that which affects the steelwork; they can vanish without a trace once management stops taking an interest in them.

A safety audit had been carried out on the plant and a number of weaknesses in design and operating procedures identified, but it seems that not all the recommendations had been carried out. All audits should be followed up.

The plant was half-owned by a US company—Union Carbide—and half-owned locally. In the case of such joint ventures it is important to make clear who is responsible for safety in both design and operations.

The plant was originally built 1½ miles from the nearest housing, but a shanty town grew up next to the plant. In many countries zoning laws prevent such development, but not in India (or, if there are such laws they are ignored). The Bhopal tragedy reinforces the need for controls to prevent hazardous plants being located close to residential areas and to prevent residential areas being built close to hazardous plants.

The press reports gave the impression that emergency planning was not as good as it might have been. This is another aspect to which industry and government should, and in many countries do, pay attention.

Bhopal has made many people ask if engineers, particularly chemical engineers, receive sufficient training in safety and loss prevention as students and when they join industry as designers or as members of the operating team. In the United Kingdom the Institution of Chemical Engineers requires all new members to have followed a course of study which includes safety and loss prevention (2), but in many countries, including the United States, the majority of chemical engineers get no such training as students. Many companies have in-house programs for the training of new staff in safety and loss prevention, but in others there is little or no systematic training. Perhaps this book will help to spread knowledge of good practice, and of accidents that have occurred, among both students and those already in industry.

However, perhaps the most important lesson to be learned from the Bhopal disaster is the need to develop plant designs which use less hazardous raw materials or not so many of the hazardous ones. Most plants are made safe by adding protective equipment, which may fail or prove inadequate, as at Bhopal. In contrast, if we do not have hazardous material present, it cannot leak out. "What you don't have, can't leak." If we do not have so much hazardous material present, leaks matter less. Such plants are said to be "inherently safer." Reference 3 describes this philosophy in more detail and gives many examples where plants have been or could be made safer by reducing inventories. The concept of inherent safety was developed after Flixborough (see Section 2.4) but progress has been slower than was hoped. Most chemical engineers have continued to try to *control* hazards by adding to their plants large quantities of protective hardware. Perhaps now we shall see more interest in ways of *avoiding* hazards by inherently safer designs. Applying this philosophy to Bhopal leads us to ask what research, if any, had been done to find routes to the final product that did not involve the production of hazardous intermediates and if it was really necessary to carry such large intermediate stocks.

MEXICO CITY

Two weeks before the disaster at Bhopal, on November 19, 1984, there was a huge fire and explosion at a liquefied petroleum gas processing plant and distribution center in Mexico City, Mexico. Over 550 people were killed, over 2,000 injured and 10,000 made homeless; 350,000 were evacuated from their homes. At the time of writing the initial cause of the disaster is not known, but according to press reports it started with a leak from a tank truck. The leak ignited (the source of ignition was said to be the flare stack) and the fire heated nearby storage vessels which BLEVEd, that is, the metal became softened and the vessels burst (See Section 8.1). It was reported that altogether twelve vessels burst, releasing their contents and adding to the conflagration. It seems that the plant was not laid out in a way that prevented the spread of fire and, in addition, as at Bhopal, a shanty town had grown up next to the plant. The disaster reinforces the need for the precautions described in Chapter 8.

SAO PAULO

1984 was a bad year for the oil and chemical industries. In another incident, on February 25, 1984, in Cubatao in Sao Paulo, Brazil, at least 508 people, most of them young children, were killed when a 2-ft diameter gasoline pipe ruptured and 700 tons of gasoline spread across a strip of swamp. The incident received little publicity, but it seems that once again a shanty town had been built illegally, on stilts over the swamp. The cause of the pipe rupture was not reported, though it was said to have been brought up to pressure in error and it was also stated that there was no way of monitoring the pressure in the pipeline (4).

REFERENCES

1. F. P. Lees, *Loss Prevention in the Process Industries,* Vol. 2, Butterworths, 1980, pp. 887, 898.
2. *A Scheme for a Degree Course in Chemical Engineering,* Institution of Chemical Engineers, Rugby, UK, 1983.
3. T. A. Kletz, *Cheaper, Safer Plants or Wealth and Safety at Work,* Institution of Chemical Engineers, Rugby, UK, 1984.
4. *Hazardous Cargo Bulletin,* June 1984 p. 34.

Recommended Reading

Descriptions of other case histories can be found in the following publications.

1. F.P. Lees, *Loss Prevention in the Process Industries,* Butterworths, 1980, Vol. 2, Appendix 1.
2. *Fire Protection Manual for Hydrocarbon Processing Plants,* edited by C.H. Vervalin, Vol. 1, Third Edition, 1985; Vol. 2, 1981, Gulf Publishing Co., Houston.
3. *Hazard Workshops Modules,* Institution of Chemical Engineers, Rugby, UK. The notes are supplemented by slides.
4. *Loss Prevention Bulletin,* published every 2 months by the Institution of Chemical Engineers, Rugby, UK.
5. *Safety Digests of Lessons Learned,* Volumes 2-5, American Petroleum Institute, 1979-1981.
6. *Hazard of Water, Hazard of Air, Safe Furnace Firing,* etc., nine booklets published by The American Oil Company. Now out-of-print but worth searching for in libraries.
7. *IP Safety News*—an occasional insert inside *Petroleum Review,* Institute of Petroleum, London.
8. Chemical Manufacturers Assn., Washington, D.C. *Case Histories.* No new ones are being published but bound volumes of old ones are available. They are, however, rather brief.

Index